Disease Epidemiology in Farm Animal Production

Disease Epidemiology in Farm Animal Production

Guest Editors

Gianmarco Ferrara
Serena Montagnaro

MDPI

Basel • Beijing • Wuhan • Barcelona • Belgrade • Novi Sad • Cluj • Manchester

Guest Editors

Gianmarco Ferrara
Department of Veterinary
Medicine and Animal
Productions
University of Naples
Naples
Italy

Serena Montagnaro
Department of Veterinary
Medicine and Animal
Productions
University of Naples
Naples
Italy

Editorial Office
MDPI AG
Grosspeteranlage 5
4052 Basel, Switzerland

This is a reprint of the Special Issue, published open access by the journal *Animals* (ISSN 2076-2615), freely accessible at: https://www.mdpi.com/journal/animals/special_issues/K6B1B9M6GU.

For citation purposes, cite each article independently as indicated on the article page online and as indicated below:

Lastname, A.A.; Lastname, B.B. Article Title. *Journal Name* **Year**, *Volume Number*, Page Range.

ISBN 978-3-7258-6760-8 (Hbk)
ISBN 978-3-7258-6761-5 (PDF)
https://doi.org/10.3390/books978-3-7258-6761-5

Cover image courtesy of Gianmarco Ferrara

Contents

Preface

Every day, hundreds of diseases pose a danger to livestock farms. Some have a direct influence; others are more subtle, and their impact is noticeable over time. Some impose limits and obligatory health measures. Some affect only animals of certain ages, while others affect the entire herd. Some are new and emerging, while others are old, but no less serious. Regardless of the impact, there is just one watchword: knowledge as a way to prevent!

Gianmarco Ferrara and Serena Montagnaro
Guest Editors

animals

MDPI

Article

Positivity Status and Molecular Characterization of Porcine Parvoviruses 1 Through 8 (PPV1-PPV8) from Slaughtered Pigs in China

Dashi Zhao [1,†], Hong Lin [1,2,3,4,5,†], Zitao Huang [6], Yajie Zhou [1], Wenhao Qi [1], Meng Cui [1], Ming Qiu [1], Jianzhong Zhu [1,2,4,5] and Nanhua Chen [1,2,3,4,5,*]

1 College of Veterinary Medicine, Yangzhou University, Yangzhou 225009, China; 18246986065@163.com (D.Z.); linhonglynn@yzu.edu.cn (H.L.); zhouyj202405@163.com (Y.Z.); qiwenhao2829@163.com (W.Q.); cuimeng201902@163.com (M.C.); qiuming1997@126.com (M.Q.); jzzhu@yzu.edu.cn (J.Z.)

2 Joint International Research Laboratory of Agriculture and Agri-Product Safety, Yangzhou 225009, China

3 International Research Laboratory of Prevention and Control of Important Animal Infectious Diseases and Zoonotic Diseases of Jiangsu Higher Education Institutions, Yangzhou 225009, China

4 Jiangsu Co-Innovation Center for Prevention and Control of Important Animal Infectious Diseases and Zoonoses, Yangzhou University, Yangzhou 225009, China

5 Comparative Medicine Research Institute, Yangzhou University, Yangzhou 225009, China

6 Animal Health Supervision Institute of Fengxi District, Chaozhou 521031, China; 13450818935@163.com

* Correspondence: chnhlh@126.com

† These authors contribute equally to this work.

Simple Summary: In addition to porcine parvovirus 1 (PPV1), seven new PPVs (PPV2-PPV8) have been identified in the last two decades. However, the prevalence and evolution of PPVs in slaughtered pigs in China are still unclear. The infection status in slaughtered pigs could reflect the overall health situation during pig production in swine herds. Therefore, we detected the infection status of PPVs in 353 samples collected from slaughtered pigs in six regions of China in 2023. Overall, 79.32% of the samples were PPV positive, with 67.50% PPV-positive samples co-infecting with two to six PPVs. Six species of PPVs were detected, except for PPV4 and PPV8. Representative PPV genomes were determined for evaluating evolutionary relationships and detecting recombination events. A genome-based phylogenetic tree confirmed the PCR results, and cross-over events were detected in the PPV2 and PPV3 strains identified in this study. This study provided the first clue on the prevalence and evolution of PPVs in slaughtered pigs in China.

Abstract: Porcine parvoviruses one through eight (PPV1-PPV8) are prevalent in Chinese swine herds. However, the infection status of all these PPVs in slaughtered pigs is still unclarified. In this study, we detected PPV1-PPV8 in 353 tissue samples collected from slaughtered pigs from six regions of China in 2023. At least one species of PPV was detected in 79.32% of the samples (280 out of 353). Six PPV species were detected, except for PPV4 and PPV8, in slaughtered pigs, within which PPV3 (49.86%), PPV2 (42.49%), and PPV7 (42.21%) were predominant, followed by PPV1 (13.31%), PPV6 (13.31%), and PPV5 (8.22%). Noticeably, co-infection was frequently detected, with 67.50% of PPV-positive samples (189 out of 280) co-infecting with two to six PPVs. In addition, one representative genome for each detected PPV was determined. Multiple sequence alignment determined a large number of substitutions in capsid proteins of PPVs. Genome-based phylogenetic analysis confirmed the PCR detection results. Recombination detection identified two potential recombinants (PPV2 GDCZ2023-2088 strain and PPV3 HLJSYS2023-1654 strain) in slaughtered pigs. Overall, this study provides new insights into the prevalence and evolution of PPVs, particularly in slaughtered pigs in China.

Keywords: PPV1-PPV8; prevalence; co-infection; genome; slaughtered pigs

1. Introduction

Parvoviruses (PVs) are small non-enveloped DNA viruses containing a linear single-stranded genome, which may infect a wide range of animals [1]. In mammals, PVs are closely associated with nervous and respiratory diseases in humans [2], hepatitis in horses [3], enteritis in dogs [4], panleukopenia in cats [5], and reproductive diseases in pigs [6]. The genomes of porcine parvoviruses (PPVs) are about 4.0 ~ 6.3 kb in length, containing two open reading frames (ORF1 and ORF2). ORF1 encodes a nonstructural protein, whilst ORF2 encodes a capsid protein [7]. Eight PPVs (PPV1 to PPV8) have been detected in wild boars and domestic pig herds [8–10], which can be divided into four genera of parvoviruses [7,11]. In detail, PPV1 and PPV8 belong to *Protoparvovirus*, PPV2 and PPV3 are grouped in *Tetraparvovirus*, PPV4, PPV5, and PPV6 are clustered within *Copiparvovirus*, and PPV7 is grouped within *Chaphamaparvovirus*. PPV1 was first isolated in Germany in 1965 [12]. PPV2 was unexpectedly amplified from swine sera in Myanmar in 2001 [13]. PPV3 (porcine hokovirus) was first identified in Hong Kong in 2008 [14]. PPV4 was detected in a diseased pig co-infected with PCV2 in the United States in 2010 [15]. PPV5 was identified in the United States in 2013 [16]. PPV6 was found in aborted pig fetuses in China in 2014 [17]. PPV7 was first described in healthy adult pigs in the United States in 2016 [18]. PPV8 was identified in PRRSV-positive samples in China in 2021 [11]. PPVs are widely spread in wild boars all around the world, including Serbia, Italy, Spain, Turkey, Romania, South Korea, and China [7,9,10,19]. They are generally detected as the most common pathogens in wild boars. For instance, two recent studies showed that PPVs could be detected in 56% and 44.4% of samples from Serbia and Italy, respectively [9,10]. More importantly, all PPVs could be detected from clinically healthy and diseased domestic pigs. The clinical symptoms may include fever, rash, dyspnea, porcine dermatitis, and nephropathy syndrome (PDNS), while the potential pathological lesions include lung consolidation, lymph node hemorrhage, kidney pathology, liver pathology, pericardium pathology, enteric pathology, meningitis, and lymphadenopathy [7]. PPV1 is the major agent causing SMEDI syndrome (stillbirths, mummification, embryonic death, and infertility) [7]. Even though the pathogenicity of new PPVs (PPV2-PPV8) is still unclarified, they have been proposed as potential pathogens of porcine respiratory disease complex (PRDC) [20]. Although there are PPV commercial vaccines available, PPV infection is still not well controlled worldwide. Therefore, it is important to monitor the prevalence and evolution of PPVs in pig populations.

PPV1 was first identified in China in 1983 [21]. In the last two decades, all new PPVs (PPV2 to PPV8) have also been detected in Chinese swine herds [7,11,14,17]. In addition, previous studies showed that PPVs can be detected at every stage of the pig production cycle in different types of samples (such as lung and lymph nodes) [7,22,23]. A recent study detected PPVs in the gilts showing that new PPVs (such as PPV4 and PPV6) might affect reproductive performance [23]. Noticeably, PPVs are more prevalent in finishing and nursery pigs than in suckling pigs [7]. The infection status in slaughtered pigs could reflect the overall health situation during pig production in swine herds. However, the infection status of PPVs in slaughtered pigs has not been determined yet. Moreover, mutation and recombination are two key mechanisms for virus evolution. Mutations in PPVs are potentially associated with variations in the pathogenicity and protective immune responses [24,25]. Recombination events are also frequently detected in different PPVs, including PPV1, PPV2, PPV3, and PPV7 [7,26]. However, the mutation and recombination situations of PPVs in slaughtered pigs were still unknown. In addition, PPVs have been detected in different types of tissues, indicating that they have broad tropism [27]. However, they were more commonly found in lung and lymphoid tissues such as lymph nodes, tonsils, and spleens [8], Therefore, we evaluated the infection status of PPVs in slaughtered pigs using 353 tissue samples (lungs and lymph nodes) collected from six regions of China in 2023. Furthermore, representative PPV-positive samples were submitted to complete genome sequencing, multiple sequence alignment, genome-based phylogenetic analysis, and recombination detection.

2. Materials and Methods

2.1. Sample Collection

A total of 353 tissue samples (including 338 lungs and 15 lymph nodes) from slaughtered pigs were submitted from six regions (Heilongjiang, Shandong, Sichuan, Henan, Guangdong, and Beijing) of China to Yangzhou University from February 18th to December 10th in 2023. These samples were mailed from slaughterhouses from distinct regions of China. However, detailed information about these slaughterhouses or animal owners was not provided. Therefore, we only record the regions where the samples came from. Even though neither the slaughterhouse owners nor the animal owners provided the hard copy consents to give us the permissions of sample collection, they, by default, allowed us to use these samples by submitting them to our laboratory for potential viral detection.

2.2. PPV Detection

The infection status of PPVs (PPV1-PPV8) in slaughtered pigs was determined by PCR assays described previously [7,11]. Briefly, total DNAs were extracted from tissue samples using the HiPure Tissue DNA minikit (Magen, Guangzhou, China). Viral DNAs were eluted using 50 μL nuclease-free double distilled water (ddH_2O) and stored at −40 °C until used. The concentrations of primer pairs and amplification conditions were optimized accordingly [7]. Each of the PCR assays was performed in a 20 μL reaction system containing 2 μL DNA, 0.5 μL corresponding primer pair (10 μM), 7.5 μL ddH_2O, and 10 μL Premix Taq (TaKaRa, Dalian, China). The amplification was carried out at 35 cycles of 98 °C 10 s, 55 °C 30 s, and 72 °C for 1 min according to the manufacturer's instructions. The PCR products were detected in 1.0% agarose gel electrophoresis with a 1× TAE buffer. The obtained amplicons were sent out for Sanger sequencing to confirm the PCR results (Genewiz, Suzhou, China).

2.3. PPV Genome Sequencing

Representative PPV-positive samples were used for complete genome sequencing with primers shown in our previous study [7]. The primers could amplify overlapped fragments covering the entire PPV genome. Each fragment was triply sequenced to ensure the accuracy of each sequence. The obtained sequences were assembled using DNAMAN 6.0 software, and six nearly complete PPV genomes were obtained in this study.

2.4. Multiple Alignment and Phylogenetic Analysis

To estimate the evolutionary relationships between PPV genomes obtained from slaughtered pigs in this study and other PPV genomes from the GenBank database, the obtained six PPV genomes and forty representative PPV genomes (5 for each PPV) were aligned by the ClustalX 2.1 (University College Dublin, Dublin, Ireland) [28,29]. And, then, a genome-based phylogenetic tree was constructed with MEGA 6.06 (Tokyo Metropolitan University, Tokyo, Japan) [7,30]. The phylogenetic tree was built using the 46 aligned sequences by the neighbor-joining method and the maximum composite likelihood model, including transitions and transversions, substitutions, homogeneous patterns among lineages, and uniform rates among sites. In addition, the complete deletion option to treat gaps and missing data were selected. The robustness was estimated by bootstrapping with 1000 replicates.

2.5. Recombination Detection

To determine the intraspecies recombination events during the generation of PPV strains from slaughtered pigs, all available PPV genomes (45 PPV1, 121 PPV2, 49 PPV3, 79 PPV5, 105 PPV6, and 145 PPV7), except for PPV4 and PPV8, from the GenBank database were used for screening the cross-over events by recombination detection program 4 (RDP4) (University of Cape Town, Cape Town, South Africa) [31]. Seven methods, including RDP, GENECONV, BootScan, MaxChi, Chimaera, SiScan, and 3Seq, were utilized for recombination detection and breakpoint determination. The p value cut off was set at 0.05. $p < 0.05$ indicated that the cross-over events are significant.

3. Results

3.1. The Prevalence of PPVs in Slaughtered Pigs in China

To test the prevalence of PPVs in slaughtered pigs in 2023, 353 tissue samples collected from six regions of China were detected. As shown in Table 1 and Table S1, at least one species of PPV was detected in 79.32% of the samples (280 out of 353). Six PPVs (PPV1, PPV2, PPV3, PPV5, PPV6, and PPV7) were detected, except for PPV4 and PPV8, in slaughtered pigs, within which PPV3 (176 out of 353, 49.86%), PPV2 (150/353, 42.49%), and PPV7 (149/353, 42.21%) were predominant, followed by PPV1 (47/353, 13.31%), PPV6 (47/353, 13.31%), and PPV5 (29/353, 8.22%). In addition, all six PPVs could be detected in both lung and lymph node samples (Table S1). These results showed that PPVs are highly prevalent in slaughtered pigs in China.

Table 1. Distribution of PPV1–8 in 353 samples from slaughtered pigs from different regions of China in 2023.

Region	No.	PPV1	PPV2	PPV3	PPV4	PPV5	PPV6	PPV7	PPV8
Heilongjiang	31 *	+	+	+	-	-	+	+	-
Shandong	37	+	+	+	-	+	+	+	-
Sichuan	34	-	+	+	-	+	+	+	-
Henan	70	+	+	+	-	+	+	+	-
Guangdong	167	+	+	+	-	+	+	+	-
Beijing	14	+	+	+	-	+	+	+	-
Total	353	47/353 (13.31%)	150/353 (42.49%)	176/353 (49.86%)	0/353 (0%)	29/353 (8.22%)	47/353 (13.31%)	149/353 (42.21%)	0/353 (0%)

* The number indicates sample numbers collected from each region.

3.2. Co-Infection Status of PPVs in Slaughtered Pigs

To clarify the co-infection status of PPVs in slaughtered pigs, the positivity of each sample was determined. As shown in Table 2, simplex, duplex, triplex, quadruplex, quintuplex, and sextuplex PPV infections were detected in 91, 102, 59, 17, 8, and 3 tissue samples from slaughtered pigs. In PPV-positive samples, even though 32.50% of the samples (91 out of 280) were infected by one species of PPV, co-infection was frequently detected, with 67.50% of the samples (189 out of 280) co-infecting with two to six types of PPVs. In addition, similar co-infection status of PPVs could be detected in both lung and lymph node samples (Table S1). These results supported that co-infection among different species of PPVs commonly occurred in slaughtered pigs in China.

Table 2. Infection and co-infection of distinct PPVs in clinical samples.

Infection Status	Numbers	Types
Simplex infection	91	PPV1 (11) *, PPV2 (16), PPV3 (28), PPV5 (1), PPV6 (2), PPV7 (33)
Duplex infection	102	PPV1+2 (3), PPV1+3 (9), PPV1+6 (1), PPV1+7 (3), PPV2+3 (35), PPV2+5 (2), PPV2+6 (1), PPV2+7 (16), PPV3+6 (6), PPV3+7 (19), PPV5+6 (1), PPV5+7 (4),PPV6+7 (2)
Triplex infection	59	PPV1+2+3 (4), PPV1+2+7 (2), PPV1+3+5 (1), PPV1+3+6 (1), PPV1+3+7 (1), PPV1+5+7 (1), PPV2+3+5 (2), PPV2+3+6 (4), PPV2+3+7 (33), PPV2+5+6 (1), PPV2+5+7 (2), PPV2+6+7 (2), PPV3+6+7 (5)
Quadruplex infection	17	PPV1+2+3+7 (3), PPV2+3+5+6 (2), PPV2+3+5+7 (4), PPV2+3+6+7 (7), PPV3+5+6+7 (1)
Quintuplex infection	8	PPV1+2+3+6+7 (4), PPV2+3+5+6+7 (4)
Sextuplex infection	3	PPV1+2+3+5+6+7 (3)
Septuplex infection	0	/ #
Octuplex infection	0	/

* The numbers in the brackets are the sample numbers of each type of infection status. # The diagonal indicates none.

3.3. Evolutionary Relationship Evaluation

To estimate the molecular characteristics and evolutionary relationships between PPVs identified in this study and other Chinese PPVs, six PPV genomes (one for each species of PPV detected positive in this study) from slaughtered pigs were determined, as shown previously [7]. The obtained nearly complete PPV genomes were deposited into the GenBank database with accession numbers PQ328182-PQ328187 (Table 3). The blast results showed that our PPV genomes shared high similarity with other PPV genomes in the GenBank database. The PPV1 GDCZ2023-2622 strain shared the highest homology (99.81%) with the PPV1KUIP22-4 isolate (GenBank No. OP377056) and >99.00% similarity with other PPV1 strains. The PPV2 GDCZ2023-2088 strain showed the highest homology (99.75%) with the HuB21-2016 isolate (MN326157) and >96.66% similarity with other PPV2 genomes. The PPV3 HLJSYS2023-1654 strain shared the highest homology (98.48%) with the HBTS20180519-151 strain (MZ577031) and >96.89% similarity with other PPV3 strains. The PPV5 SCNJ2023-1865 strain showed the highest homology (99.70%) with the SDWF20170530-67 strain (MZ577037) and >99.18% similarity with other PPV5 genomes. The PPV6 GDCZ2023-2439 strain shared the highest homology (98.37%) with the SC strain (KF999684) and >97.81% similarity with other PPV6 strains. The PPV7 HNZMD2023-1903 strain showed the highest homology (95.43%) with the GX49 strain (NC_040562) and >93.35% similarity with other PPV7 genomes. The genome-based phylogenetic tree showed that our PPV genomes obtained in this study were grouped together with corresponding PPV genomes (Figure 1). The phylogenetic results not only confirmed the accuracy of the PCR results but also supported the close evolutionary relationship between PPVs in slaughtered pigs and PPVs from other production stages of pigs.

Table 3. Six PPV genomes from slaughtered pigs determined in this study.

Species	Name	Region *	Collection time	GenBank No.
PPV1	GDCZ2023-2622	Chaozhou, Guangdong	11 November 2023	PQ328182
PPV2	GDCZ2023-2088	Chaozhou, Guangdong	23 April 2023	PQ328183
PPV3	HLJSYS2023-1654	Shuangyashan, Heilongjiang	27 February 2023	PQ328184
PPV5	SCNJ2023-1865	Neijiang, Sichuan	1 March 2023	PQ328185
PPV6	GDCZ2023-2439	Chaozhou, Guangdong	10 September 2023	PQ328186
PPV7	HNZMD2023-1903	Zhumadian, Henan	2 March 2023	PQ328187

* The region indicated the city and province from where the sample was collected.

3.4. Recombination Events and Substitutions

To evaluate the role of recombination in the generation of PPVs from slaughtered pigs identified in this study, we aligned our PPV genomes with all available corresponding PPV genomes in GenBank and submitted them to intraspecies recombination analyses. PPV2 GDCZ2023-2088 and PPV3 HLJSYS2023-1654 strains were detected as recombinants by all seven methods in RDP4 (Table 4). The PPV2 GDCZ2023-2088 strain was recombined from SDWF20171225-112 and GD6-2017 viruses (Figure 2A), while PPV3 HLJSYS2023-1654 was generated by a recombination event between GD202206-4 and SD202203-3 viruses (Figure 2B). In addition, the alignments of PPV capsid proteins identified a large number of substitutions in our PPVs when compared with other Chinese PPVs (Figure S1). These results supported that finishing pigs (slaughtered pigs) could also serve as an important host for the evolution of PPVs in China.

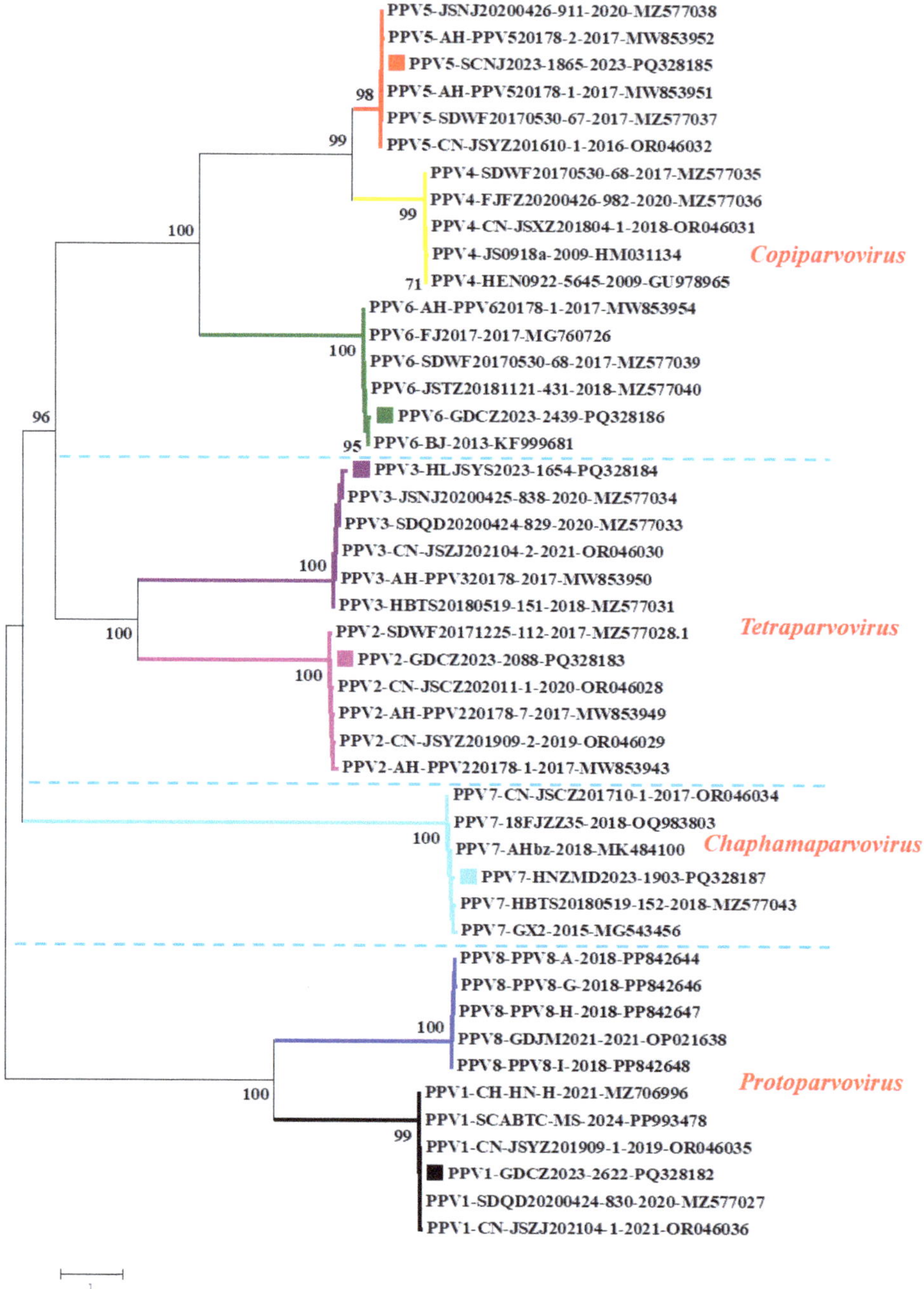

Figure 1. Genome-based phylogenetic analysis. The phylogenetic tree was constructed based on six PPV genomes obtained in this study and forty representative PPV genomes (five for each PPV) from GenBank. Distinct PPVs are shown in different colors. PPVs are clustered within four genera. Our PPV strains are highlighted with colored squares. Each virus is presented by species, virus name, year of identification, and GenBank accession number. Bootstrap values from 1000 replications are shown in each node.

Table 4. Cross-over events identified by RDP4 in this study.

Species		PPV2	PPV3
Recombinant Virus		GDCZ2023-2088	HLJSYS2023-1654
Parental viruses	Major	SDWF20171225-112	GD202206-4
	Minor	GD6-2017	SD202203-3
Breakpoints [a]	Begin	1669	1167
	End	3052	2083
Score for the seven detection methods embedded in RDP4 [b]	RDP	2.3×10^{-4}	6.2×10^{-10}
	GENECONV	4.0×10^{-11}	3.2×10^{-7}
	BootScan	2.5×10^{-16}	6.2×10^{-10}
	MaxChi	1.2×10^{-5}	8.8×10^{-9}
	Chimaera	6.4×10^{-6}	6.2×10^{-7}
	SiScan	1.6×10^{-11}	8.4×10^{-22}
	3Seq	1.9×10^{-11}	7.2×10^{-10}

[a] The breakpoints are based on the locations in the genome of the PPV2 GDCZ2023-2088 and PPV3 HLJSYS2023-1654 strains. [b] The *p* value cut off is set at 0.05. $p < 0.05$ indicates that the cross-over events are significant.

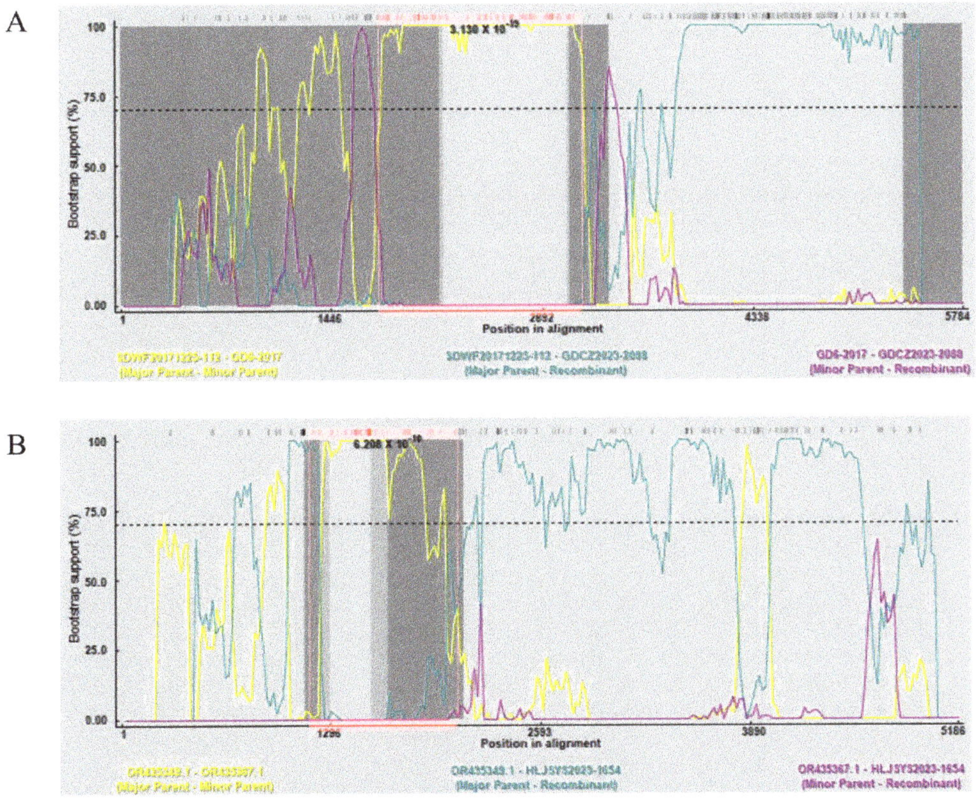

Figure 2. Potential recombination events detected in our PPV2 and PPV3 strains by RDP4. (**A**) The PPV2 GDCZ2023-2088 strain was recombined from the major parental virus SDWF20171225-112 and the minor parental virus GD6-2017. (**B**) The PPV3 HLJSYS2023-1654 strain was recombined from parental GD202206-4 and SD202203-3 viruses. The *p* value identified by the BootScan method for each cross-over event was also shown.

4. Discussion

All PPVs have been widely spread in Chinese swine herds. The majority of previous studies explored the infection status of one specific PPV or different PPVs in a specific region or specific farms [11,14,17,21,32–34]. However, very few studies focused on the prevalence and evolution of PPVs in a specific stage of the pig production cycle. In this study, we evaluated the infection status and evolution of PPVs in slaughtered pigs in China in 2023. Considering that PPVs have broad tissue tropism, they have been detected in a large number of tissue samples, including the lung, liver, brain, kidney, spleen, heart, thymus, and thyroid [27]. PPV1 has a tropism to macrophages that could migrate to the placenta and then infect the fetus [35]. PPV2-PPV8 can be detected in serum and tissues, mainly in lung and lymphoid tissues [8]. Therefore, lung and lymph node samples were collected and used for PPV detection in this study. Our results showed that several PPVs are highly prevalent in slaughtered pigs in China. In addition, mutation and recombination analyses showed that finishing pigs (slaughtered pigs) may also serve as a critical host for PPV evolution.

The prevalence of distinct PPVs in China was extensively studied. The percentages of PPV1- to PPV8-positive samples might range from 0 to 75% [11,17,32,36–38]. However, the infection status of PPVs in a specific pig production stage was rarely evaluated. Replacement gilts play an essential role in the reproductive cycle. PPV infection in gilts might affect reproductive performance. A recent study on the gilts showed that PPVs are highly prevalent in gilts, with PPV3 (40.1%), PPV5 (20.5%), PPV6 (17%), and PPV1 (14.5%) serving as prevalent viruses [23]. PPV infection in slaughtered pigs could reflect the overall PPV infection status in swine herds. Our results showed that PPVs are also highly prevalent in slaughtered pigs in China, with PPV3 (49.86%), PPV2 (42.49%), and PPV7 (42.21%) being the predominant viruses. Even though PPV4 and PPV8 were not detected in our tissue samples from slaughtered pigs in 2023, the exact infection statuses of PPV4 and PPV8 in slaughtered pigs required further investigation using more spatiotemporal representative samples.

The co-infections among distinct PPVs and PPVs with other pathogens, such as PCV2 and PRRSV, commonly occurred [7,8,23]. Viral pathogens, such as PPV1, PCV2, and PRRSV, and bacterial pathogens, including *Mycoplasma hyopneumoniae,* are all major pathogens for porcine reproductive failure and PRDC [8]. Even though the specific role of each pathogen in PRDC is still unclear, it is well known that co-infection of PPVs with PCV2 and PRRSV could influence the severity of clinical diseases [7]. In addition, new PPVs, such as PPV2, have also been associated with PRDC [20]. Therefore, not only the co-infection of PPVs with other pathogens but also the co-infection among distinct PPVs might play synergistic roles in PPV pathogenicity. In this study, co-infections among PPV1, PPV2, PPV3, PPV5, PPV6, and PPV7 were frequently detected in slaughtered pigs. However, due to the unknown healthy condition of these slaughtered pigs and no cell lines that could be used to isolate new PPVs, it is still a huge challenge to explore the influence of PPV infection and co-infection on the pathogenicity. In light of the complexity of co-infections among distinct pathogens in pigs, overall biosafety prevention and control strategies in a farm level rather than individual pig or individual disease treatments must be preferential.

To evaluate the evolutionary relationships between Chinese PPVs detected in slaughtered pigs and other Chinese PPVs, six representative PPV genomes were determined and submitted to genome-based phylogenetic analysis. Our PPVs shared high genomic similarities (>93.35%) with corresponding PPVs and had close evolutionary relationships with corresponding PPVs, suggesting that there was no significant genetic difference between PPVs in slaughtered pigs and PPVs in other stages of the pig production cycle.

Mutation is one of the major mechanisms for viral evolution. Five mutations (I215T, D378G, H383Q, S436P, and R565K) in the capsid protein of the PPV1 Kresse strain are potentially associated with the pathogenicity [24]. Substitutions at 378, 383, and 565 residues in the 3-fold spike region might also influence immune response [25,39]. Antisera from pigs infected by various PPV1 strains had high neutralizing activities against homologous PPV1 strains but low neutralizing activities against heterologous viruses [25]. A large number of substitutions were identified in our PPVs from slaughtered pigs. Whether these mutated PPVs would change their pathogenicity or affect protective immune responses required further investigation.

In addition to mutation, recombination also plays a critical role in the generation of viral genomic diversity. Recombination events are frequently detected in parvoviruses, including interspecies recombination in rodent parvoviruses and intraspecies recombination in porcine parvoviruses [26]. Intraspecies recombination events have been detected in PPV1, PPV2, PPV3, and PPV7, while no cross-over events were reported in PPV4, PPV5, PPV6, and PPV8 [7,8,26]. PPV1 2074-7 and 225b isolates were detected as potential recombinants [7,8,40]. Strong recombination signals were detected in nine PPV2 strains (F3-12R, F7-1BV, WB-102R2, WB-826MR, WB-763S, WB-804D, WB-720I, F1-23M, and F4-44M) and three PPV3 strains (F2-47M, WB-RO-369, and WB-RO-834) both in domestic pigs and wild boars [41]. Two PPV7 strains (KF4 from South Korea and HBTZ20180519-152 from China) were detected as recombinants originating from wild boars [7]. In this study, the PPV2 GDCZ2023-2088 recombinant was recombined from two PPV2 strains from domestic pigs, while the PPV3 HLJSYS2023-1654 recombinant was generated by two PPV3 strains with unclarified origin. Both this study and previous reports confirmed that PPV recombination events could be detected within and between PPVs from the ages of pigs, within and between regions/countries, and within and between strains originating from domestic pigs and wild boars [8,37]. Overall, both the mutations and recombination events identified in our PPVs from slaughtered pigs confirmed that PPVs keep evolving in finishing pigs (slaughtered pigs) in China.

This is the first study to focus on the prevalence and evolution of PPVs in slaughtered pigs in China. The results from this study supported that PPVs persistently infect pigs, even at the late stage of the pig's production cycle. In addition, the evolution of PPVs would not stop, even when the host pig was slaughtered. These findings emphasize the importance of biosecurity during the entire pig production cycle. However, only a small number of tissue samples were detected in this study. In addition, these samples were only collected from six regions of China within one year (2023). Due to these limitations, the results from this study might not reflect the exact infection status of PPVs in slaughtered pigs in China. Therefore, more spatially and temporally representative samples should be collected and evaluated in the near future.

5. Conclusions

This study provided the first clue on the infection status of PPVs in slaughtered pigs in China. In addition, our results also confirmed that finishing pigs (slaughtered pigs) also serve as a non-negligible host for PPV evolution.

Supplementary Materials: The following supporting information can be downloaded at https://www.mdpi.com/article/10.3390/ani14223238/s1, Table S1. Clinical sample infection status. The detailed infection and co-infection information of PPVs in 353 tissue samples collected from six regions of China in 2023 are shown. Figure S1. Multiple alignments of capsid proteins. (A–F) The capsid proteins from our PPVs were compared with capsid proteins from five corresponding representative PPVs.

Author Contributions: Conceptualization, N.C.; methodology, D.Z. and H.L.; software, N.C. and J.Z.; validation, N.C., DZ., H.L., Z.H., Y.Z., W.Q., M.C. and M.Q.; formal analysis, N.C., H.L., D.Z., Z.H., Y.Z., W.Q., M.C. and M.Q.; investigation, D.Z., H.L., Z.H., Y.Z., W.Q., M.C. and M.Q.; resources, N.C., Z.H. and H.L.; data curation, D.Z., H.L., Z.H., Y.Z., W.Q., M.C. and M.Q.; writing—original draft preparation, N.C. and H.L.; writing—review and editing, N.C., J.Z., D.Z. and H.L.; visualization, N.C., D.Z. and H.L.; supervision, N.C.; project administration, N.C.; funding acquisition, N.C. All authors have read and agreed to the published version of the manuscript.

Funding: This research is funded by the National Key R&D Program of China (2023YFD1800504), the Open Project Program of the International Research Laboratory of Prevention and Control of

Important Animal Infectious Diseases and Zoonotic Diseases of Jiangsu Higher Education Institutions (No. 3), the Priority Academic Program Development of Jiangsu Higher Education Institutions (PAPD), and 111 Project D18007. The APC was funded by Priority Academic Program Development of Jiangsu Higher Education Institutions (PAPD). Dashi Zhao was supported by the Science and Innovation Program of Yangzhou University (XCX20230740).

Institutional Review Board Statement: The samples were collected from dead animals and an official exemption letter was gained from the Animal Welfare and Ethics Committee, Yangzhou University.

Informed Consent Statement: Not applicable.

Data Availability Statement: The obtained nearly complete PPV genomes have been submitted to GenBank with accession numbers PQ328182-PQ328187.

Acknowledgments: The authors would like to thank Shubin Li, Binghui Feng, Chen Li, Yuejia Qiu, and Shuai Yang for their assistance in clinical sample collections.

Conflicts of Interest: The authors declare no conflicts of interest.

References

1. Toolan, H.W. The parvoviruses. *Prog. Exp. Tumor Res.* **1972**, *16*, 410–425. [PubMed]
2. Anderson, M.J.; Higgins, P.G.; Davis, L.R.; Willman, J.S.; Jones, S.E.; Kidd, I.M.; Pattison, J.R.; Tyrrell, D.A. Experimental parvoviral infection in humans. *J. Infect. Dis.* **1985**, *152*, 257–265. [CrossRef] [PubMed]
3. Ramsauer, A.S.; Badenhorst, M.; Cavalleri, J.V. Equine parvovirus hepatitis. *Equine Vet. J.* **2021**, *53*, 886–894. [CrossRef] [PubMed]
4. Mazzaferro, E.M. Update on Canine Parvoviral Enteritis. *Vet. Clin. N. Am. Small Anim. Pract.* **2020**, *50*, 1307–1325. [CrossRef]
5. Stuetzer, B.; Hartmann, K. Feline parvovirus infection and associated diseases. *Vet. J.* **2014**, *201*, 150–155. [CrossRef]
6. Redman, D.R.; Bohl, E.H.; Ferguson, L.C. Porcine parvovirus: Natural and experimental infections of the porcine fetus and prevalence in mature swine. *Infect. Immun.* **1974**, *10*, 718–723. [CrossRef]
7. Li, J.; Xiao, Y.; Qiu, M.; Li, X.; Li, S.; Lin, H.; Li, X.; Zhu, J.; Chen, N. A Systematic Investigation Unveils High Coinfection Status of Porcine Parvovirus Types 1 through 7 in China from 2016 to 2020. *Microbiol. Spectr.* **2021**, *9*, e0129421. [CrossRef]
8. Vargas-Bermudez, D.S.; Mogollon, J.D.; Franco-Rodriguez, C.; Jaime, J. The Novel Porcine Parvoviruses: Current State of Knowledge and Their Possible Implications in Clinical Syndromes in Pigs. *Viruses* **2023**, *15*, 2398. [CrossRef]
9. Ferrara, G.; Piscopo, N.; Pagnini, U.; Esposito, L.; Montagnaro, S. Detection of selected pathogens in reproductive tissues of wild boars in the Campania region, southern Italy. *Acta Vet. Scand.* **2024**, *66*, 9. [CrossRef]
10. Jezdimirovic, N.; Savic, B.; Milovanovic, B.; Glisic, D.; Ninkovic, M.; Kureljusic, J.; Maletic, J.; Radojkovic, J.A.; Kasagic, D.; Milicevic, V. Molecular Detection of Porcine Cytomegalovirus, Porcine Parvovirus, Aujeszky Disease Virus and Porcine Reproductive and Respiratory Syndrome Virus in Wild Boars Hunted in Serbia during 2023. *Vet. Sci.* **2024**, *11*, 249. [CrossRef]
11. Guo, Y.; Yan, G.; Chen, S.; Han, H.; Li, J.; Zhang, H.; Luo, S.; Liu, M.; Wu, Q.; Li, Q.; et al. Identification and genomic characterization of a novel porcine parvovirus in China. *Front. Vet. Sci.* **2022**, *9*, 1009103. [CrossRef] [PubMed]
12. Cartwright, S.F.; Lucas, M.; Huck, R.A. A small haemagglutinating porcine DNA virus. I. Isolation and properties. *J. Comp. Pathol.* **1969**, *79*, 371–377. [CrossRef] [PubMed]
13. Hijikata, M.; Abe, K.; Win, K.M.; Shimizu, Y.K.; Keicho, N.; Yoshikura, H. Identification of new parvovirus DNA sequence in swine sera from Myanmar. *Jpn. J. Infect. Dis.* **2001**, *54*, 244–245. [PubMed]
14. Lau, S.K.P.; Woo, P.C.Y.; Tse, H.; Fu, C.T.Y.; Au, W.K.; Chen, X.C.; Tsoi, H.W.; Tsang, T.H.F.; Chan, J.S.Y.; Tsang, D.N.C.; et al. Identification of novel porcine and bovine parvoviruses closely related to human parvovirus 4. *J. Gen. Virol.* **2008**, *89 Pt 8*, 1840–1848. [CrossRef] [PubMed]
15. Cheung, A.K.; Wu, G.; Wang, D.; Bayles, D.O.; Lager, K.M.; Vincent, A.L. Identification and molecular cloning of a novel porcine parvovirus. *Arch. Virol.* **2010**, *155*, 801–806. [CrossRef]
16. Xiao, C.T.; Gimenez-Lirola, L.G.; Jiang, Y.H.; Halbur, P.G.; Opriessnig, T. Characterization of a novel porcine parvovirus tentatively designated PPV5. *PLoS ONE* **2013**, *8*, e65312. [CrossRef]
17. Ni, J.; Qiao, C.; Han, X.; Han, T.; Kang, W.; Zi, Z.; Cao, Z.; Zhai, X.; Cai, X. Identification and genomic characterization of a novel porcine parvovirus (PPV6) in China. *Virol. J.* **2014**, *11*, 203. [CrossRef]
18. Palinski, R.M.; Mitra, N.; Hause, B.M. Discovery of a novel Parvovirinae virus, porcine parvovirus 7, by metagenomic sequencing of porcine rectal swabs. *Virus Genes* **2016**, *52*, 564–567. [CrossRef]
19. Park, G.N.; Song, S.; Cha, R.M.; Choe, S.; Shin, J.; Kim, S.Y.; Hyun, B.H.; Park, B.K.; An, D.J. Genetic analysis of porcine parvoviruses detected in South Korean wild boars. *Arch. Virol.* **2021**, *166*, 2249–2254. [CrossRef]
20. Nelsen, A.; Lin, C.M.; Hause, B.M. Porcine Parvovirus 2 Is Predominantly Associated With Macrophages in Porcine Respiratory Disease Complex. *Front. Vet. Sci.* **2021**, *8*, 726884. [CrossRef]
21. Hao, X.; Lu, Z.; Sun, P.; Fu, Y.; Cao, Y.; Li, P.; Bai, X.; Bao, H.; Xie, B.; Chen, Y.; et al. Phylogenetic analysis of porcine parvoviruses from swine samples in China. *Virol. J.* **2011**, *8*, 320. [CrossRef] [PubMed]

22. Igriczi, B.; Denes, L.; Schonhardt, K.; Balka, G. First Report of Porcine Parvovirus 8 in Europe: Widespread Detection and Genetic Characterization on Commercial Pig Farms in Hungary and Slovakia. *Animals* **2024**, *14*, 1974. [CrossRef] [PubMed]
23. Vargas-Bermudez, D.S.; Diaz, A.; Polo, G.; Mogollon, J.D.; Jaime, J. Infection and Coinfection of Porcine-Selected Viruses (PPV1 to PPV8, PCV2 to PCV4, and PRRSV) in Gilts and Their Associations with Reproductive Performance. *Vet. Sci.* **2024**, *11*, 185. [CrossRef] [PubMed]
24. Bergeron, J.; Hebert, B.; Tijssen, P. Genome organization of the Kresse strain of porcine parvovirus: Identification of the allotropic determinant and comparison with those of NADL-2 and field isolates. *J. Virol.* **1996**, *70*, 2508–2515. [CrossRef]
25. Zeeuw, E.J.; Leinecker, N.; Herwig, V.; Selbitz, H.J.; Truyen, U. Study of the virulence and cross-neutralization capability of recent porcine parvovirus field isolates and vaccine viruses in experimentally infected pregnant gilts. *J. Gen. Virol.* **2007**, *88 Pt 2*, 420–427. [CrossRef]
26. Shackelton, L.A.; Hoelzer, K.; Parrish, C.R.; Holmes, E.C. Comparative analysis reveals frequent recombination in the parvoviruses. *J. Gen. Virol.* **2007**, *88 Pt 12*, 3294–3301. [CrossRef]
27. Oraveerakul, K.; Choi, C.S.; Molitor, T.W. Tissue tropisms of porcine parvovirus in swine. *Arch. Virol.* **1993**, *130*, 377–389. [CrossRef]
28. Thompson, J.D.; Gibson, T.J.; Higgins, D.G. Multiple sequence alignment using ClustalW and ClustalX. *Curr. Protoc. Bioinform.* **2002**, *Chapter 2*, Unit 2.3. [CrossRef]
29. Larkin, M.A.; Blackshields, G.; Brown, N.P.; Chenna, R.; McGettigan, P.A.; McWilliam, H.; Valentin, F.; Wallace, I.M.; Wilm, A.; Lopez, R.; et al. Clustal W and Clustal X version 2.0. *Bioinformatics* **2007**, *23*, 2947–2948. [CrossRef]
30. Tamura, K.; Stecher, G.; Peterson, D.; Filipski, A.; Kumar, S. MEGA6: Molecular Evolutionary Genetics Analysis version 6.0. *Mol. Biol. Evol.* **2013**, *30*, 2725–2729. [CrossRef]
31. Martin, D.P.; Murrell, B.; Golden, M.; Khoosal, A.; Muhire, B. RDP4: Detection and analysis of recombination patterns in virus genomes. *Virus Evol.* **2015**, *1*, vev003. [CrossRef] [PubMed]
32. Huang, L.; Zhai, S.L.; Cheung, A.K.; Zhang, H.B.; Long, J.X.; Yuan, S.S. Detection of a novel porcine parvovirus, PPV4, in Chinese swine herds. *Virol. J.* **2010**, *7*, 333. [CrossRef] [PubMed]
33. Wu, R.; Wen, Y.; Huang, X.; Wen, X.; Yan, Q.; Huang, Y.; Ma, X.; Cao, S. First complete genomic characterization of a porcine parvovirus 5 isolate from China. *Arch. Virol.* **2014**, *159*, 1533–1536. [CrossRef] [PubMed]
34. Wang, Y.; Yang, K.K.; Wang, J.; Wang, X.P.; Zhao, L.; Sun, P.; Li, Y.D. Detection and molecular characterization of novel porcine parvovirus 7 in Anhui province from Central-Eastern China. *Infect. Genet. Evol.* **2019**, *71*, 31–35. [CrossRef]
35. Mengeling, W.L.; Lager, K.M.; Vorwald, A.C. The effect of porcine parvovirus and porcine reproductive and respiratory syndrome virus on porcine reproductive performance. *Anim. Reprod. Sci.* **2000**, *60–61*, 199–210. [CrossRef]
36. Sun, J.H.; Huang, L.P.; Wei, Y.W.; Wang, Y.P.; Chen, D.J.; Du, W.J.; Wu, H.L.; Liu, C.M. Prevalence of emerging porcine parvoviruses and their co-infections with porcine circovirus type 2 in China. *Arch. Virol.* **2015**, *160*, 1339–1344. [CrossRef]
37. Cadar, D.; Lorincz, M.; Kiss, T.; Novosel, D.; Podgorska, K.; Becskei, Z.; Tuboly, T.; Csagola, A. Emerging novel porcine parvoviruses in Europe: Origin, evolution, phylodynamics and phylogeography. *J. Gen. Virol.* **2013**, *94 Pt 10*, 2330–2337. [CrossRef]
38. Xing, X.; Zhou, H.; Tong, L.; Chen, Y.; Sun, Y.; Wang, H.; Zhang, G. First identification of porcine parvovirus 7 in China. *Arch. Virol.* **2018**, *163*, 209–213. [CrossRef]
39. Streck, A.F.; Canal, C.W.; Truyen, U. Molecular epidemiology and evolution of porcine parvoviruses. *Infect. Genet. Evol.* **2015**, *36*, 300–306. [CrossRef]
40. Leal, E.; Villanova, F.E.; Lin, W.; Hu, F.; Liu, Q.; Liu, Y.; Cui, S. Interclade recombination in porcine parvovirus strains. *J. Gen. Virol.* **2012**, *93 Pt 12*, 2692–2704. [CrossRef]
41. Cadar, D.; Csagola, A.; Kiss, T.; Tuboly, T. Capsid protein evolution and comparative phylogeny of novel porcine parvoviruses. *Mol. Phylogenet Evol.* **2013**, *66*, 243–253. [CrossRef]

animals

MDPI

Article

Serological and Molecular Investigation of Infectious Laryngotracheitis Virus in Chickens from Robe Town, Southeastern Ethiopia

Samuel Abebe [1], Gianmarco Ferrara [2], Belayneh Getachew [3], Eyob Hirpa [4] and Nebyou Moje [4,*]

1 Faculty of Veterinary Medicine, Hawassa University, Hawassa P.O. Box 05, Ethiopia; edegdvm2016@gmail.com
2 Department of Veterinary Science, University of Messina, 98166 Messina, Italy; gianmarco.ferrara@unina.it
3 National Veterinary Institute (NVI), Bishoftu P.O. Box 19, Ethiopia; belaynehgetachew@gmail.com
4 College of Veterinary Medicine and Agriculture, Addis Ababa University, Bishoftu P.O. Box 34, Ethiopia; eyob.hirpa@aau.edu.et
* Correspondence: nebhawas@gmail.com or nebyou.moje@aau.edu.et

Simple Summary: Infectious laryngotracheitis virus (ILTV) causes avian infectious laryngotracheitis (ILT), a highly contagious acute respiratory disease that affects chickens. This infection is ubiquitous globally and decreases poultry production, with significantly more catastrophic repercussions in states with unstable economic systems. In this study, a total of 240 sera from eight farms (including commercial and backyard) were sampled to evaluate ILTV exposure. A total of 64 samples tested positive by commercial ELISA. The risk analysis identified higher prevalences in backyard chickens and farms that introduced new animals from other farms. Furthermore, 15 suspect animals were sampled for the viral isolation in embryonated eggs. Isolation was successful in six samples, of which four were confirmed by PCR using specific primers. This study highlights the presence of this virus in different types of poultry farms in Southeastern Ethiopia and identifies some management practices that favor the spread of this infection.

Abstract: Infectious laryngotracheitis virus (ILTV) is responsible for avian infectious laryngotracheitis (ILT), a highly contagious acute respiratory disease affecting chickens. However, there is limited information on ILTV and its distribution in Ethiopia, particularly in the southeastern region. The aim of this study was to establish the serological prevalence and molecular evidence in commercial and backyard chickens from Robe town, Southeastern Ethiopia. A cross-sectional study was conducted between December 2021 and June 2022, collecting 240 serum samples from randomly selected chickens belonging to eight kebeles (farms) using systematic random sampling. ILTV-specific antibodies were detected using a commercial indirect enzyme-linked immunosorbent assay (ELISA). From 240 serum samples, 26.7% were positive for ILTV antibodies. Logistic regression analysis identified the type of poultry farm (backyard) and the introduction of chickens from other farms as potential risk factors associated with ILTV exposure. Tracheal tissue and oropharyngeal and tracheal swabs were collected from suspected chickens for isolation and molecular detection. A total of six samples were successfully isolated in embryonated eggs (40%), with four of them verified with a specific PCR. These findings documented the presence of ILTV in the study area, which needs further insight to fully understand the actual spread of ILTV and quantify the damage caused to the poultry sector.

Keywords: avian infectious laryngotracheitis; ILTV; infectious laryngotracheitis virus; molecular characterization; seroprevalence; risk factors; Southeastern Ethiopia

1. Introduction

Chicken production is widely spread in Ethiopia and represents a valuable source of protein and income, especially for rural areas [1]. About 97% of the Ethiopian poultry population consists of indigenous chickens, while the remaining 3% consists of imported exotic and hybrid breeds of chickens [2]. The consequences of globalization, climate change, and the rapidly expanding poultry population favor the emergence of several diseases [3]. Among these emerging diseases, avian infectious laryngotracheitis (ILT) is an acute, highly transmissible viral disease of chickens, mainly affecting the upper respiratory tract. The etiological agent is *Gallid alphaherpesvirus* type 1 (GaHV-1), which belongs to the genus Iltovirus, family Herpesviridae, and subfamily Alphaherpesvirinae [4,5]. The virus has a linear double-stranded DNA genome of approximately 155 kb that encodes 80 viral proteins [6]. These envelope proteins, including gB, gC, gD, gG, gH, gJ, gM, and gN, are glycosylated, and they are deputies of several functions, such as the mediation of attachment and entry into the host cell and the interaction with the host immune system [7,8].

Although chickens are the primary host, occasional infections of pheasants, partridges, and peafowl have been reported, while several species, including starlings, sparrows, crows, pigeons, and ducks, seem to be resistant to the virus [3,9]. The virus is horizontally transmitted, and the primary replication site is the tracheal mucosa [10]. The outcome of infection depends on the virulence of the strain or co-infection with other respiratory pathogens (avian influenza virus (AIV), Newcastle disease virus (NDV), and infectious bronchitis virus (IBV)), with mortality rates ranging from 5% to 70% [11–14]. Two distinct clinical presentations are reported (severe and milder forms). The severe form causes significant dyspnea, expectoration of bloody mucus, and sneezing [3,4]. The milder form is associated with conjunctivitis, mucoid tracheitis, sinusitis, swollen infraorbital (almond-shaped eyes), nasal discharge, reduced egg production, poor weight gain, and low mortality [3,15]. GaHV-1, like other members of the herpes virus family, induces latent infections due to its persistence in the trigeminal ganglion of the central nervous system after 7 days of acute infection (which can reactivate under stress conditions) [15,16].

A laboratory diagnosis is required for ILT since other diseases cause similar clinical signs and lesions. Several methods can be carried out to confirm ILTV, including histopathology to detect syncytia and intranuclear inclusion bodies (INIBs), virus isolation, antigen detection with immunofluorescent antibodies (IFA) or immunohistochemistry (IHC), enzyme-linked immunosorbent assay (ELISA), direct electron microscopy (DEM), and DNA detection methods [17]. Furthermore, methods have developed rapidly in recent years. These can identify ILTV quickly and accurately, are highly sensitive, and successfully identify ILTV in clinical samples including the trachea, larynx, and conjunctiva [18,19].

Knowledge about the spread of this pathogen is fundamental in areas such as Ethiopia, where subsistence farming is also practiced. The first report of ILTV was reported by Mekibib et al. [4] from the southern part of Ethiopia. Furthermore, while other reports exist [20–23], there is a critical lack of information in Southeastern Ethiopia. The aim of this study was to determine the seroprevalence of ILTV in commercial and backyard farms from Robe town, Southeastern Ethiopia, and evaluate the potential risk factors involved in the spread of this infection. A second aim of this study was to attempt viral isolation to establish the molecular prevalence and strain for future investigations.

2. Materials and Methods

2.1. Description of the Study Area

The study was conducted in Robe Town, West Bale zone, Oromia regional state, Southeastern Ethiopia, from December 2021 to June 2022 (Figure 1). The area has an average annual temperature and humidity of 16.5 °C and 64%, respectively. The agro-climatic condition of the area is highland. In the area, there are two rainy seasons: the first and main season extends from August to December, and the second and shorter rainy season is from April to July. The dry season covers December to March [24]. In the study area, a total of 152,189 poultry are raised [25]. Robe Town and its surroundings were chosen because of the presence of traditional small poultry farms as well as the import of several exotic breeds of poultry from Central Ethiopia to the area, which may have contributed to the emergence of ILT.

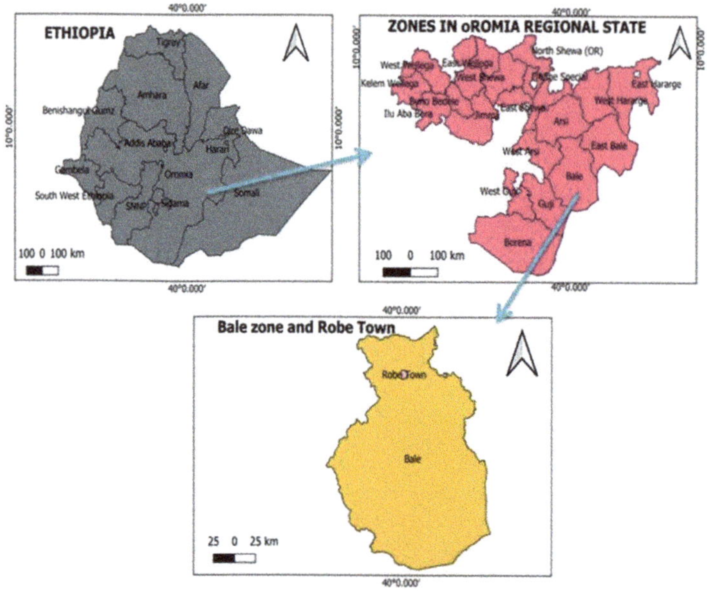

Figure 1. Map of the study district (Developed by QGIS).

2.2. Study Design, Study Population, and Sampling

A cross-sectional study was conducted from December 2021 to June 2022. The study populations included indigenous and exotic breeds of chickens reared in eight backyard and commercial farms in Robe Town. The chickens included in the study were healthy (for serological analysis) or diseased (for the viral isolation attempts), greater than 8 weeks old, and of both sexes. No commercial ILTV vaccine was available, and no vaccination programs had been implemented in Ethiopia. A multistage sampling was implemented to select the study zone, and a systematic random sampling technique was employed to select the village, flocks, and the number of chickens to sample from each farm. Purposive sampling through the evaluation of health status by clinical examination was employed to select diseased chickens (dyspnea, expectoration of bloody mucus, sneezing, high mortality, conjunctivitis, sinusitis, swollen infraorbital sinuses, nasal discharge, reduced egg production, poor weight gain) for isolation and molecular detection of ILTV. The sample size required for the seroprevalence study was determined based on sample size determination in random sampling for an infinite population with an expected prevalence of 19.4% [21] and with a confidence level of 95% and 5% desired absolute precision [26]. A total of 240 sera were collected from commercial and backyard chickens.

$$n = \frac{1.96^2 + Pexp(1 - Pexp)}{d^2} = \frac{1.96^2 + 0.194(1 - 0.194)}{0.05^2} = 240$$

where n = sample size; P_{exp} = expected prevalence; and d = desired absolute precision.

The equal interval estimation during systematic random sampling was described as follows:

$$\text{jth} = \text{N/n}$$

where N indicates the total population of poultry and n indicates the sample size:

$$\text{jth} = \text{N/n} = 152{,}189/240 = 12\text{th}$$

Since maternal immunity is expected to develop during the first three weeks of life, chickens less than three weeks of age were excluded from the study. Each sample was collected aseptically from the wing vein (about 2–3 mL of blood) using a sterile syringe with 21-gauge needles and a vacutainer tube. Blood samples were immediately transported to the laboratory, centrifuged, and stored at −20 °C before being processed. Clinical samples (tracheal/oropharyngeal swabs and tracheal tissue) were taken from ILT-suspected chickens, inserted into cryovial tubes containing a virus transport medium supplemented with antibiotics, transported to the laboratory, and stored at −80 °C until the laboratory analysis was performed. Each sample was accompanied by a questionnaire with relevant information related to each chicken, including location, age, breed, sex, production type, and rearing method.

All chickens were sampled according to international animal care and use guidelines adopted by the Research Ethical Committee (ARSEC) of NVI [27]. Ethical clearance for the study was provided by the Research Ethics Committee of the Faculty of Veterinary Medicine, Hawassa University (FVM, HU). The research ethics committee of the FVM-HU reviewed and discussed this research on 19 September 2021 (Reference No. 620/w, date 7 July 2022).

2.3. *Laboratory Analysis*

2.3.1. Serological Analysis

Serological tests were performed by an indirect commercial ELISA (ILTV Antibody Test Kit IDvet® Screen® ILT Indirect, 310 rue Louis Pasteur, Grabels, France) to measure specific antibodies against GaHV-1 in chicken sera. The test was performed according to the manufacturer's instructions. Briefly, each test sample was diluted at 1:500 with sample diluents and incubated at room temperature for 60 min. After incubation, each well was washed three times with approximately 300 µL. After each wash, 100 µL of conjugate (anti-chicken IgG labeled with alkaline phosphatase) was added to each well and incubated at room temperature for 60 min. Following a further washing step, 100 µL of substrate reagent (TMB) was added to each well and incubated at room temperature for 15 min. Finally, 100 µL of stop solution (2M H_2SO_4) was added to each well to stop the reaction. The microtiter ELISA plate was placed in the ELISA reader to measure the OD (optical density) at 405 nm wavelength and interpret the results.

2.3.2. Isolation and Molecular Detection

A total of 15 samples (6 oropharyngeal swabs and 9 tracheal swabs/tissues) were collected from suspected chickens for ILT virus isolation and molecular detection. Tracheal swabs/tissues were suspended in 10% (*w*/*v*) of sterile phosphate-buffered saline solution supplemented with penicillin and streptomycin (1000 µg/mL). The suspension was transferred into a sterile centrifuge tube and centrifuged at 3000 rpm for 10 min. The supernatant was harvested and employed for virus isolation and molecular detection. The tracheal tissue sample was chopped into small pieces using a sterile scalpel blade and minced using a mortar and pestle. The specimens were inoculated onto the chorioallantoic membranes (CAMs) of 10-day-old specific-pathogen-free (SPF) chicken embryos, which were incubated at 37 °C and examined daily for 5 days. CAMs and the allantoic fluids were harvested five days post-inoculation to collect the virus [28–30]. Briefly, embryonated SPF eggs were disinfected with 70% ethanol and inoculated with 0.2 mL of 10% of the supernatant using an insulin needle. Eggs were incubated at 37 °C and checked daily for embryo mortality. Any mortality within the first 24 h post-inoculation was considered non-specific, and the eggs were discarded. The dead embryo eggs were chilled at 4 °C for 24 h then opened aseptically, and the embryos were examined for gross ILT lesions [31]. The harvested allantoic fluid was added to sterile cryovial tubes and stored at −20 °C until DNA extraction was performed.

DNA extraction from the field sample tracheal/oropharyngeal swabs, tracheal tissue suspension, and allantoic fluid (n = 15) was performed with the QIAGEN DNA Mini

Column Kit (QIAGEN, Frankfurt, Germany). Conventional PCR was used with a set of primers that specifically amplify a 688 bp fragment of the ICP4 gene. The PCR was conducted using Bio Rad 2729 Thermal Cycler (Hercules, CA, USA) in a reaction volume of 25 µL, containing 5 µL of 10× Dream Taq buffer, 2 µL RNAs free water, 5 µL of each 2 mM of deoxynucleotide triphosphate, 5 µL of Dream Taq DNA polymerase, 2 µL of 5 pm/µL Primer ILT (F: 5′ ACT GAT AGC TTT TCG TAC AGC ACG 3′ and R: 5′-CAT-CGG-GAC-ATT-CTC-CAG-GTA-GCA-3), and 3 µL template DNA and resulted in a 688 bp amplicon of the ICP4 gene fragment [32]. Thermal cycling conditions included an initial denaturation at 94 °C for 3 min, followed by 35 cycles of a three-step amplification protocol (denaturation at 94 °C for 30 s, annealing at 60 °C for 45 s, and elongation at 72 °C for 1:50 s), and finally one cycle of elongation at 72 °C for 10 min. PCR products were analyzed by 1.5% (*w*/*v*) agarose electrophoresis gel and visualized using a UV-lamp camera.

2.4. Statistical Analysis

Descriptive statistics were employed to summarize the study variables. Binary logistic regression was used to identify the potential risk factor for the ILTV. A first univariate logistic regression analysis was used, and those factors with a p-value < 0.25 were subjected to multivariable logistic regression. Odds ratios at a 95% confidence interval were used to express the strength of the association of the factors with the occurrence of the disease. Moreover, Hosmer–Lemeshow goodness of fit test was used to check the model's adequacy. In the final model, a p-value of less than 0.05 with a 95% confidence interval (CI) was used to declare the associated factors. All the statistical analyses were performed by SPSS version 28 statistical software.

3. Results

3.1. Seroprevalence and Associated Risk Factors of Infectious Laryngotracheitis Virus

In the current study, 64 out of 240 blood samples tested were positive for ILTV-specific antibodies. The overall individual seroprevalence was 26.7% (ranging from 15.8 to 36.9% among different farms), and all the sampled poultry farms resulted positive (Table 1). Univariate analysis (chi-square) was used to investigate the influence of individual and managerial risk factors on ILTV seroprevalence. Several factors were positively associated with higher seroprevalence rates, including breed (local) and type of farm (backyard). Moreover, a higher prevalence was found in chickens raised on farms that introduced animals from other farms. Production purpose, age, sex, and farm size did not affect the ILTV seroprevalence. Variables with a p-value lower than 0.25 from the univariable analysis were included in the final multivariable logistic model based on a stepwise backward elimination procedure (Table 2). Backyard chickens were 1.464 times more likely to be affected by ILTV than commercial chickens (Table 3). For a unit increase in the number of introduced chickens, the odds of being affected by ILTV were increased by 1.52. The model has a good fit since the Hosmer and Lemeshow tests could not reject the hypothesis of model appropriateness with a value of p = 0.95.

Table 1. Kebele (farm)-level seroprevalence of ILTV in chickens in the study district.

Kebele	No. Sample	Positive	Prevalence (%)	95% CI
Alage	33	12	36.4	22.19, 53.38
Basaso	19	4	21.1	8.51, 43.33
Bole	57	11	19.3	11.13, 31.34
Robe/01	19	3	15.8	5.52, 37.57
Robe/02	19	6	31.6	15.36, 53.99
Robe/03	27	7	25.9	13.17, 44.68
Robe/04	20	5	25.0	11.19, 46.87
Shallo	46	16	34.8	22.68, 49.23
Total	240	64	26.7	21.47, 32.6

Table 2. Univariable analysis for the occurrence of ILTV in chickens and its potential risk factors.

Factor	Examined (*n*)	Positive	Proportion (%)	95%CI	*p*-Value
Breed					
Exotic	226	56	24.78	9.15–30.41	
Local	14	8	57.14	31.22–83.07	**0.008**
Sex					
Male	46	9	19.57	8.10–31.03	
Female	194	55	35.71	15–43.28	0.22
Purpose					
Broiler	47	10	21.28	9.58–32.98	
Layer	193	54	27.98	21.65–34.31	0.35
Age					
>20 weeks	193	55	28.5	22.13–34.87	
8–20 weeks	47	9	19.5	7.9–30.4	0.19
Farm type					
Backyard	70	31	44.29	32.65–55.92	
Commercial	170	33	19.4	13.47–25.36	**<0.0001**
Chicken introduction					
No	17	13	76.47	53–90	
Yes	223	51	22.87	18–29	**<0.0001**

Bold *p*-values are significant ones.

Table 3. Multiple variable logistic analysis for the occurrence of ILTV and its potential risk factors.

Variable	*n*	Positive	Proportion (%)	AOR (95% CI)	*p*-Value
Chicken introduction					
No	17	13	76.47		
Yes	223	51	22.87	6.79 (2, 23.1)	**0.002**
Farm type					
Backyard	70	31	44.29		
Commercial	170	33	19.4	2.3 (1.19, 4.46)	**0.013**

Bold *p*-values are significant ones; AOR: Adjusted odds ratio.

3.2. Isolation and Molecular Detection of Infectious Laryngotracheitis Virus

Out of 15 ILTV infection-suspected samples that were inoculated onto the CAMs of 10-day-old embryonated SPF eggs via three consecutive passages, only six samples killed the embryo that showed white pock lesions on the CAM of the embryonated SPF egg (Table S1). Further analysis of molecular detections revealed that out of the 15 DNA samples tested, 4 (26.7%) were positive for ILTV (Table S2). Notably, PCR amplification produced a band of 688 bp in three tracheal swabs and one oropharyngeal swab sample (Figure 2). The positive samples were collected from layers and exotic-breed chickens.

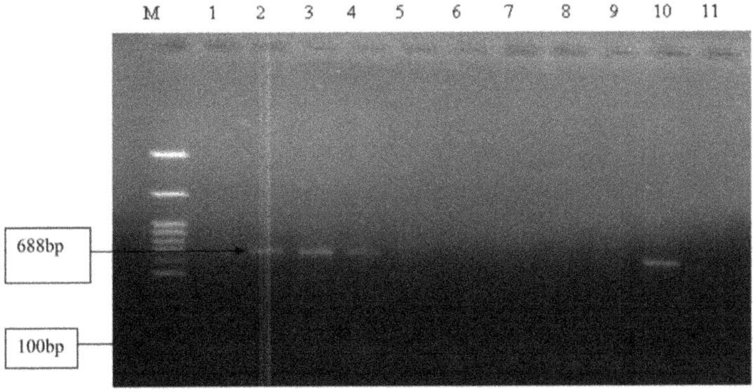

Figure 2. PCR amplification of 688 bp fragment of the ICP4 gene from ILTV-infected field samples. (M = 100-bp DNA ladder; 1 = Positive Control; 2, 3, 4, and 10 = field sample isolates showing ILTV-specific 688 bp products; 11 = Negative control).

4. Discussion

The overall seroprevalence of ILTV infection obtained from the present study across the eight kebeles was 26.7%. All sampled farms had positive animals, demonstrating the widespread diffusion of this pathogen in the studied area. This prevalence was similar to the finding of Baksi [33], who reported a 26.77% prevalence in India. However, the current finding was lower than that of the findings of Birhan et al. [23], Salhi et al. [34], Roba et al. [22], Mijanur et al. [35], Jahan et al. [36], and Shaza et al. [37], who reported a prevalence of 59.1% in Northwestern Ethiopia, 56.25% in Algeria, 54.7% in the Oromia region's Ada'a districts, and 81.47%, 92.28%, and 96.7% in Bangladesh, respectively. Some of the above-stated countries use immunization against this virus; therefore, the antibodies found could possibly be vaccine-related. On the other hand, lower exposure rates were found in Central and South Ethiopia (19.4%), Ecuador (0.19%), Finland (12%), Bangladesh (0.4–17.33%), Iran (13%), and North Central Nigeria (1.2%) [21,38–43]. Furthermore, the differences observed between studies could be due not only to different epidemiological situations but also to the type of sampling carried out, the type of test used, the study period, etc.

The risk analysis revealed, as previously observed for other infections, that backyard chickens were more exposed to ILTV, most likely due to poorer biosecurity frequently observed on family farms and more interaction with wild animals [44]. On the other hand, backyard chickens may serve as a reservoir for wild birds due to their close association. The introduction of new chickens to the farm was another risk factor that was statistically associated with increased seroprevalence. This approach increases the chance of infection with any disease, particularly herpesvirus infection. In fact, apparently healthy animals with latent diseases might be placed on the farm, and when stressed, they facilitate the spread of the virus due to viral reactivation [45,46]. The absence of a difference in prevalence between layer and broiler chickens was unexpected, as, although all chickens are susceptible to the virus, the literature reports greater risks of infection between layers due to their longer productive lives [46].

The molecular prevalence (40%) obtained from testing suspected chickens was very similar to the seroprevalence rate. Previous studies performed in Ethiopia found molecular prevalence of 0% and 11% in 2017 and 2022, respectively [47,48]. In this case, the differences in prevalence derive above all from the type of matrix used rather than from the PCR protocol, which is rather standardized. In fact, experimental infections have established that 7 days after infection, the areas with the highest viral load are the conjunctiva, the trachea, the lungs, and the spleen [49]. Embryonated eggs inoculated with the field virus also have high viral loads, increasing the reliability of molecular methods.

Furthermore, the use of embryonated eggs (Table S1) also allows for the isolation of the virus so that it can be studied, sequenced, or used for the production of vaccines. Although the present study was performed on a limited number of samples, it demonstrated the circulation of ILTV in an area of Ethiopia where it had not yet been described and identified risk factors to take into consideration for the management of this infection. Further studies are necessary to fully understand the diffusion and impact that ILTV has on Ethiopian territory.

5. Conclusions

The present study demonstrated an overall ILTV seroprevalence of 26.7% and confirmed four ILTV-positive results in the backyard and commercial chickens of Robe Town, Southeastern Ethiopia. These data suggest that ILTV is widespread among backyard and commercial poultry farms in the study area. Introduced chickens and backyard production systems had increased seropositivity for ILTV. This study provides suggestions for control and biosecurity measures in the poultry farms to eradicate ILTV and represents a baseline for further study concerning ILTV in Ethiopia (such as sequencing of the viral genome to identify the strain).

Supplementary Materials: The following supporting information can be downloaded at: https://www.mdpi.com/article/10.3390/ani14223227/s1. Table S1: Isolation of ILTV from the field suspected samples using 10 days old embryonated SPF eggs for three consecutive passages; Table S2: Detail description of the active case with results.

Author Contributions: Conceptualization, S.A. and N.M.; methodology, S.A., B.G. and N.M.; software, N.M.; validation, S.A., N.M., B.G., E.H. and G.F.; formal analysis, S.A., N.M. and G.F.; investigation, S.A., B.G. and N.M.; resources, N.M., B.G. and G.F.; data curation, N.M., G.F. and E.H.; writing—original draft preparation, N.M.; writing—review and editing; S.A., N.M., G.F. and E.H.; visualization, S.A., N.M., G.F. and B.G.; supervision, N.M. and B.G.; resources, validation, visualization, B.G. and N.M. All authors have read and agreed to the published version of the manuscript.

Funding: This research received no external funding.

Institutional Review Board Statement: The study was conducted in accordance with the Declaration of Helsinki and approved by the Institutional Review Board (or Ethics Committee) of Hawassa University, Faculty of Veterinary Medicine, on 19 September 2021 (Reference No. 620/w, date 7 July 2022).

Informed Consent Statement: Informed consent was obtained from all animal owners involved in the study.

Data Availability Statement: All data regarding this paper are available in this publication. However, further information can be requested from the corresponding author.

Acknowledgments: We would like to acknowledge all poultry/chicken owners for allowing us to get samples from their animals. National Veterinary Institute (NVI) management teams, and specifically Takele Abayneh, director of the institute are highly appreciated for giving their permission to allow us to use their laboratory resources. NVI laboratory personnels are also appreciated for their role. All personnel that had a role in the laboratory processing are appreciated.

Conflicts of Interest: The authors declare no conflicts of interest.

References

1. Tadelle, D.; Million, T.; Alemu, Y.; Peters, K. Village chicken production systems in Ethiopia: Use patterns and performance valuation and chicken products and socioeconomic functions of chicken. *Livest. Res. Rural. Dev.* **2003**, *15*, 1–9. Available online: https://www.cabidigitallibrary.org/doi/full/10.5555/20033132152 (accessed on 15 February 2023).
2. CSA. *Agricultural Sample Survey. Report on Livestock and Livestock Characteristics. The Federal Democratic Republic of Ethiopia, Private Peasant Holdings*; Statistical Bulletin 570; Central Statistical: Addis Ababa, Ethiopia, 2013.
3. Bagust, T.J.; Jones, R.C.; Guy, J.S. Avian infectious laryngotracheitis. *Rev. Sci. Tech.* **2000**, *19*, 483–492. [CrossRef] [PubMed]
4. Ou, S.C.; Giambrone, J.J.; Macklin, K.S. Comparison of a TaqMan real-time polymerase chain reaction assay with a loop-mediated isothermal amplification assay for detection of *Gallid herpesvirus 1*. *J. Vet. Diagn. Investig.* **2012**, *24*, 138–141. [CrossRef] [PubMed]
5. Jaisree, S.; Bharathi, R.; Shoba, K.; Ramesh, A.; Vijayarani, K.; Vairamuthu, S. Detection and Molecular Characterization of Avian Infectious Laryngotracheitis Virus Isolated from a Breeder Flock. *J Anim. Res.* **2021**, *11*, 257–262. [CrossRef]
6. Menendez, K.R.; Garcia, M.; Spatz, S.; Tablante, N.L. Molecular epidemiology of infectious laryngotracheitis: A review. *Avian Pathol.* **2014**, *43*, 108–117. [CrossRef]
7. Fuchs, W.; Veits, J.; Helferich, D.; Granzow, H.; Teifke, P.J.; Mettenleiter, C.T. Molecular biology of avian infectious laryngotracheitis virus. *Vet. Res.* **2007**, *38*, 261–279. [CrossRef]
8. Kongsuwan, K.; Johnson, A.; Sheppard, M. Identification of an infectious laryngotracheitis virus gene encoding an immunogenic protein with a predicted M_r of 32 kilodaltons. *Virus Res.* **1993**, *29*, 125–140. [CrossRef] [PubMed]
9. Saif, Y.M.; Glisson, J.R.; Fadly, A.M.; Mcdougald, L.R.; Nolan, L.K.; Swayne, D.E. Infectious Laringotracheitis. In *Diseases of Poultry*, 13th ed.; Garcıa, M., Spatz, S., Guy, J.S., Eds.; Iowa State Press: Ames, IA, USA, 2013; pp. 161–179.
10. Coppo, M.J.C.; Hartley, C.A.; Devlin, J.M. Immune responses to infectious laryngotracheitis virus. *Dev. Comp. Immunol.* **2013**, *41*, 454–462. [CrossRef]
11. Swayne, D.E.; Boulianne, M.; Logue, C.M.; McDougald, L.R.; Nair, V.; Suarez, D.L. Infectious Laryngotracheitis. In *Diseases of Poultry*, 14th ed.; García, M., Spatz, S., Eds.; Wiley-Blackwell: Hoboken, NJ, USA, 2020; pp. 189–203.
12. Wolfrum, N. Infectious laryngotracheitis: An update on current approaches for prevention of an old disease. *J. Anim. Sci.* **2020**, *98*, S27–S35. [CrossRef]
13. Saif, Y.M.; Glisson, J.R.; Fadly, A.M.; Mcdougald, L.R.; Nolan, L.K.; Swayne, D.E. Laryngotracheitis. In *Diseases of Poultry*, 12th ed.; Guy, J.S., Garcia, M., Eds.; Blackwell Publishing: Arres, IA, USA, 2008; pp. 137–152.
14. Addis, B.; Tadesse, D.; Mekuriaw, S. Study on major causes of chicken mortality and associated risk factors in Bahir Dar Zuria District, Ethiopia. *Afr. J. Agric. Res.* **2014**, *9*, 3465–3472. [CrossRef]

15. Rojs, O.Z.; Dov, A.; UrosKrapez, U.; Zlabravec, Z.; Racnik, J.; Slavec, B. Detection of Laryngotracheitis Virus in Poultry Flocks with Respiratory Disorders in Slovenia. *Viruses* **2021**, *13*, 707. [CrossRef] [PubMed]
16. Ferrara, G.; Iovane, V.; Moje, N.; Improda, E.; Iovane, G.; Pagnini, U.; Montagnaro, S. Cattle exposure to bubaline herpesvirus (BuHV-1) in Southern Italy: A hidden threat for IBR eradication? *Prev. Vet. Med.* **2024**, *224*, 106116. [CrossRef] [PubMed]
17. Crespo, R.; Woolcock, P.R.; Chin, R.P.; Shivaprasad, H.L.; García, M. Comparison of diagnostic techniques in an outbreak of infectious laryngotracheitis from meat chickens. *Avian Dis.* **2007**, *51*, 858–862. [CrossRef] [PubMed]
18. Zhao, Y.; Kong, C.; Cui, X.; Cui, H.; Shi, X.; Zhang, X.; Hu, S.; Hao, L.; Wang, Y. Detection of infectious laryngotracheitis virus by real-time PCR in naturally and experimentally infected chickens. *PLoS ONE* **2013**, *8*, e67598. [CrossRef] [PubMed]
19. Kammon, A.; Shabba, J.; Abouzeed, Y.; Asheg, A. Isolation, Serological and Molecular Detection of Infectious Laryngotracheitis Virus (ILTV) in Chickens in Libya. *Approaches Poult. Dairy Vet. Sci.* **2020**, *7*, 655–665. [CrossRef]
20. Mekibib, B.; Dejene, S.; Abebe, R. First Report of Avian Infectious Laryngotracheitis Outbreak in Small Scale Chicken Flocks Around Hawassa City, Ethiopia. *Anim. Health Prod.* **2019**, *67*, 239–245.
21. Tesfaye, A.; Sahle, M.; Sori, T.; Kassa, T.; Garoma, A.; Koran, T.; Dima, C.; Guyassa, C.; Hilu, H.; Guta, S.; et al. Infectious Laryngotracheitis Virus in Commercial and Backyard Chicken Production Systems in Central and South Ethiopia (first report) ILT in Ethiopian Poultry Production. *J. Appl. Poult. Res.* **2019**, *28*, 1324–1329. [CrossRef]
22. Roba, Y.T.; Tadesse, D.; Assefa, Z.; Tesfaye, A. Seroprevalence of infectious laryngotracheitis disease in backyard chickens in villages of Ada'a district, Oromia, Ethiopia: First report. *Trop. Anim. Health Prod.* **2020**, *52*, 3109–3112. [CrossRef]
23. Birhan, M.; Syoum, A.; Ibrahim, S.M.; Fentahun, T.; Mohammed, A.; Berhane, N.; Bitew, M.; Gelaye, E.; Atanaw, M.B.; Getachew, B.; et al. Serological Evidence of Infectious Laryngotracheitis Infection and Associated Risk Factors in Chickens in Northwestern Ethiopia. *Sci. World J.* **2022**, *2022*, 6096981. [CrossRef]
24. SARC. Profile of Sinana Agricultural Research Center (SARC). In *Working Document Series 1, November 2001;* Oromia Agricultural Research Institute: Addis Ababa, Ethiopia, 2001.
25. CSA. *The 2001/02 Ethiopian Agricultural Sample Enumeration (ESAE)*; Farm Practice 2001–2002 (1994 E.C); Central Statistical Authority (CSA): Addis Ababa, Ethiopia, 2004.
26. Thrusfield, M. *Veterinary Epidemiology*, 4th ed.; John Wiley and Sons, Inc.: Hoboken, NJ, USA, 2018; pp. 272–293.
27. OIE. Collection and Shipment of Diagnostic Specimens. In *Terrestrial Manual 2008;* World Organisation for Animal Health: Paris, France, 2008; pp. 1–14.
28. OIE. Chapter 3.3.3.—Avian infectious laryngotracheitisAvian infectious laryngotracheitis. In *WOAH Terrestrial Manual;* World Organisation for Animal Health: Paris, France, 2021; Available online: https://www.woah.org/fileadmin/Home/fr/Health_standards/tahm/3.03.03_AVIAN_INF_LARYNGO.pdf (accessed on 15 May 2023).
29. Ebrahimi, M.M.; Shahsavandi, S.; Yousefi, A.R.; Ebrahimi, N. Isolation, identification and chemical inactivation of infectious laryngotracheitis virus for use as a vaccine candidate. *Acta Virol.* **2021**, *65*, 33–41. [CrossRef]
30. Yi, C.; Li, G.; Mu, Y.; Cui, S.; Zhang, D.; Xu, Q.; Liang, C.; Wang, M.; Zhou, S.; Zhou, H.; et al. Isolation, identification, molecular and pathogenicity characteristics of an infectious laryngotracheitis virus from Hubei province, China. *Poult. Sci.* **2024**, *103*, 103271. [CrossRef] [PubMed]
31. OIE. Avian infectious laryngotracheitis. In *OIE Terrestrial Manual;* World Organisation for Animal Health: Paris, France, 2014; pp. 1–11.
32. Chacon, J.L.; Ferreira, A.J.P. Differentiation of field isolates and vaccine strains of infectious laryngotracheitis virus by DNA sequencing. *Vaccine* **2009**, *27*, 6731–6738. [CrossRef] [PubMed]
33. Baksi, S. Sero-prevalence of infectious laryngotracheitis of poultry in India. *Indian J. Poult. Sci.* **2016**, *51*, 234–236. [CrossRef]
34. Salhi, O.; Messaï, C.R.; Ouchene, N.; Boussaadi, I.; Kentouche, H.; Kaidi, R.; Khelef, D. Indicators and risk factors of infectious laryngotracheitis in layer hen flocks in Algeria. *Vet. World* **2021**, *14*, 182–189. [CrossRef] [PubMed]
35. Rahman, M.; Uddin, K.; Hassan, Z.; Rahman, M.; Ali, Z.; Rahman, L.; Akter, M.R.; Rahman, M. Seroprevalence study of infectious laryngotracheitis virus antibody of commercial layer in Gazipur Districts of Bangladesh. *Asian J. Med. Biol. Res.* **2018**, *4*, 1–6. [CrossRef]
36. Jahan, M.S.; Khan, R.F.; Nazir, H.N.; Amin, M.M.; Rahman, M.B. Sero-Surveillance of Infectious Laryngotracheitis in Layer Birds in Bangladesh. *Microbes Health* **2012**, *1*, 38–40. [CrossRef]
37. Mutwakil, S.H.; Ballal, M.; Osman, O.H.; Ballal, A.; Osman, O.H. Infectious Laryngotracheitis Virus Antibodies in some States of the Sudan. *Sudan J. Vet. Res.* **2011**, *26*, 71–76. Available online: https://www.sudanjvr.net/journal/1727.pdf (accessed on 10 May 2023).
38. Pohjola, L.; Tammiranta, N.; Ek-Kommonen, C.; Soveri, T.; Hänninen, M.L.; Ahomaa, M.F.; Huovilainen, A. A survey for selected avian viral pathogens in backyard chicken farms in Finland. *Avian Pathol.* **2017**, *46*, 166–172. [CrossRef]
39. Bhuiyan, Z.; Ali, Z.; Moula, M.; Bary, A.; Arefin, N.; Giasuddin; Khan, Z. Sero-prevalence of major avian respiratory diseases in broiler and sonali chicken in selected areas of Bangladesh. *J. Adv. Vet. Anim. Res.* **2019**, *6*, 561–566. [CrossRef]
40. GhalyanchiLangeroudi, A.; Hosseini, H.; Fallah, M.H.; Aghaeean, L.; Dizaji, R.E.; Ziafati, Z.; Modiri, A.; Almasi, Y.; BaGholamian, B.; Ashouri, A.; et al. Original article serological survey of infectious laryngotracheitis in broiler flocks, Iran, 2018. *Iran. J. Virol.* **2020**, *14*, 1–5. Available online: http://journal.isv.org.ir/article-1-378-en.html (accessed on 20 March 2023).

41. Uddin, I.; Sen, B.; Islam, S.; Islam, M.S.; Das, S.; Sultana, N.; Ripa, R.N.; Kashem, A.; Kamaruddin, K.M. Seroepidemiology of infectious laryngotracheitis (ILT) in the commercial layer farms of Chittagong district, Bangladesh. *Adv. Anim. Vet. Sci.* **2014**, *2*, 316–320. [CrossRef]
42. Shittu, I.; Sulaiman, K.I.; Gado, A.D.; Egbuji, A.N.; Ndahi, M.D.; Pam, E.; Joannis, T.M. Sero-epizootiological investigation of infectious laryngotracheitis infection in commercial poultry of Plateau state, north central Nigeria. *J. Immunoass. Immunochem.* **2016**, *37*, 368–375. [CrossRef] [PubMed]
43. Garrido, A.; Barrionuevo, M.; Jara, I.S.; Sandoval, P.; Alfonso, P.; Barrera, M. Serologic and molecular survey of avian infectious laryngotracheitis in Ecuador. *Ecuad. Calid.* **2016**, *3*, 43–51. [CrossRef]
44. Sonaiya, E.B.; Swan, S.E.J. *Small-Scale Poultry Production*; FAO: Rome, Italy, 2004.
45. Adene, D.F. *The Cornerstones in Poultry Health and Production: Concepts, Costs and the Contemporary Applications*; Ibadan University Press: Ibadan, Nigeria, 2007.
46. Cover, M.S. The early history of Infectious laryngotracheitis. *Avian Dis.* **1996**, *40*, 494–500. [CrossRef] [PubMed]
47. Hutton, S.; Bettridge, J.; Christley, R.; Habte, T.; Ganapathy, K. Detection of infectious bronchitis virus 793B, avian metapneumovirus, *Mycoplasma gallisepticum* and *Mycoplasma synoviae* in poultry in Ethiopia. *Trop. Anim. Health Prod.* **2017**, *49*, 31. [CrossRef]
48. Tekelemariam, T.H.; Walkden-Brown, S.; Atire, F.A.; Tefera, D.A.; Alemayehu, D.H.; Gerber, P.F. Detection of Chicken Respiratory Pathogens in Live Markets of Addis Ababa, Ethiopia, and Epidemiological Implications. *Vet. Sci.* **2022**, *9*, 503. [CrossRef]
49. Tran, T.T.; Nazir, S.; Yegoraw, A.A.; Assen, A.M.; Walkden-Brown, S.W.; Gerber, P.F. Detection of infectious laryngotracheitis virus (ILTV) in tissues and blood fractions from experimentally infected chickens using PCR and immunostaining analyses. *Res. Vet. Sci.* **2020**, *134*, 64–68. [CrossRef]

animals

MDPI

Article

Herd-Level Risk Factors Associated with *Mycoplasma bovis* Serostatus in Youngstock on Irish Dairy Farms

Marie-Claire McCarthy [1,*], Luke O'Grady [2], Conor G. McAloon [2] and John F. Mee [3]

1 Cork Regional Veterinary Laboratory, Department of Agriculture, Food and the Marine, T12 XD51 Cork, Ireland
2 School of Veterinary Medicine, University College Dublin, Belfield, D04W6F6 Dublin 4, Ireland; luke.ogrady@ucd.ie (L.O.); conor.mcaloon@ucd.ie (C.G.M.)
3 Teagasc, Animal and Bioscience Research Department, Dairy Production Research Centre, Moorepark, Fermoy, P61P302 Co Cork, Ireland; john.mee@teagasc.ie
* Correspondence: marieclaire.mccarthy@agriculture.gov.ie

Simple Summary: *Mycoplasma bovis* is a significant pathogen in cattle, causing diseases such as respiratory illness, mastitis, arthritis, and reproductive failure. First detected in Ireland in 1994, it has since become a major health issue in Irish cattle herds. This study investigated the risk factors associated with *M. bovis* seropositivity in replacement dairy heifers across 105 Irish dairy herds. Ten heifers per herd were sampled during three periods: spring 2018, spring 2019, and autumn 2019. Results showed that seropositivity varied over time, with 50.4% of herds having at least one positive heifer in spring 2018, 35.2% in spring 2019, and 45.7% in autumn 2019. Risk factors for seropositivity included the purchase of cattle, managing multiple land parcels, and housing heifer calves separately from bull calves. Shared airspace between calves and older animals also increased the risk of *M. bovis* seropositivity. Conversely, feeding more colostrum reduced the odds of seropositivity. These findings highlight the importance of implementing strong biosecurity measures, improving calf management, and enhancing colostrum feeding practices to control the spread of *M. bovis* in Irish dairy herds.

Abstract: *Mycoplasma bovis* is a globally significant pathogen of cattle associated with a wide range of clinical syndromes, including respiratory disease, mastitis, arthritis, otitis, and reproductive failure. Since its detection in Ireland in 1994, *M. bovis* has become a significant contributor to morbidity and mortality in Irish cattle. This study aimed to investigate herd-level risk factors associated with *M. bovis* seropositivity in replacement dairy heifers, using data from 105 Irish dairy herds. Ten heifers per herd were sampled on three occasions: spring 2018, spring 2019, and autumn 2019. Seropositivity was evaluated using two thresholds: $\geq$1 positive heifer (Model $\geq$ 1POS) and $\geq$3 positive heifers (Model $\geq$ 3POS). *M. bovis* seropositivity varied over time, with at least one positive heifer in 50.4% (95% confidence interval (CI): 40.5–60.3) of herds in spring 2018, 35.2% (95% CI: 26.2–45.1) in spring 2019, and 45.7% (95% CI: 36.0–55.7) in autumn 2019. Herds with three or more positive heifers increased from 31.4% (95%CI: 22.7–41.2) in spring 2018 to 42.9% (95% CI: 33.2–52.9) by autumn 2019. Risk factors for *M. bovis* seropositivity included the purchase of cattle, which significantly raised the odds of seropositivity across multiple visit periods (Model $\geq$ 1POS: Odds ratio (OR) 3.84, $p = 0.02$; Model $\geq$ 3POS: OR 3.69, $p = 0.02$). Managing more than three land parcels, housing heifer calves separately from bull calves, and sharing airspace between calves and older animals also increased seropositivity risks. Conversely, more colostrum feeds reduced the risk of seropositivity (Model $\geq$ 1POS: OR 0.81, $p = 0.05$), while colostrum quality assessment and feeding waste milk showed a trend toward increased risk. These findings suggest the importance of robust biosecurity measures, including limiting cattle purchases, improving calf management, and enhancing colostrum feeding practices, to control the spread of *M. bovis*. This study provides valuable insights into the epidemiology of *M. bovis* in Irish dairy herds, emphasising the need for targeted biosecurity and surveillance to safeguard herd productivity.

Keywords: *Mycoplasma bovis*; cattle health; dairy heifers; herd-level risk factors; dairy herds

1. Introduction

Mycoplasma bovis is a significant pathogen affecting cattle populations globally. Clinically, mycoplasmosis manifests in various forms, with arthritis and mastitis being prevalent in adult cattle, while pneumonia, arthritis, and otitis are primarily associated with infections in calves [1–3]. These infections contribute to significant economic losses in dairy and beef industries due to reduced milk production, treatment costs, and culling of infected animals [4]. For example, in the U.S., *M. bovis* respiratory infections in beef cattle, which lead to reduced weight gain and lower carcass value, are estimated to cost USD 32 million annually. Additionally, the economic losses from bovine mastitis caused by *M. bovis* may exceed those from respiratory infections, with estimates reaching up to USD 108 million per year [5]. Since its first detection in Ireland in 1994 [6], *M. bovis* has become endemic and is now considered a major cause of youngstock morbidity and mortality. For example, it has been identified as the aetiological agent in 14.3% of neonatal (0–1 month) and 13.7% of older calf (1–5 months of age) deaths attributed to respiratory disease in Ireland in 2022, respectively [7]. The herd-level prevalence of *M. bovis* in Irish dairy herds has recently been estimated at 45% using bulk tank milk (BTM) ELISA testing [8]. In that study herd seropositivity was associated with herd size, the number of contiguous farms, and geographical location. While that study provides valuable insights into *M. bovis* seroprevalence at the herd-level, research focusing specifically on Irish dairy herds—particularly regarding youngstock and replacement dairy heifers—remains limited. Additionally, using bulk tank milk as a matrix for seroprevalence studies presents several inherent limitations. While BTM ELISA testing is both cost-effective and convenient—requiring only a single pooled milk sample for large-scale herd-level screening—it has significant limitations. Most notably, it excludes non-lactating animals such as calves, heifers, sick cows, and dry cows, which may serve as substantial reservoirs of *M. bovis* infection, potentially leading to inaccurate herd-level prevalence estimates. Moreover, pooling of milk samples can dilute antibody concentrations, particularly when there are small numbers of infected cows, which reduces the test sensitivity. External factors, such as herd size, may also affect test accuracy and introduce biases in prevalence estimates. In contrast, seroprevalence studies in youngstock cohorts offer a more detailed assessment of *M. bovis* infection dynamics within a herd. These studies not only identify seroprevalence trends but also provide critical insights into the natural progression of *M. bovis* infections, such as the duration of antibody persistence following natural infection. Moreover, since adult animals may have persistent antibodies from previous infections, making it difficult to distinguish between current and past exposure, sampling youngstock avoids this issue. Their antibody response is more likely to reflect recent or ongoing infection.

In Swiss dairy herds, *M. bovis* infections were more prevalent in high-yielding herds and those with a high rate of animal movements (e.g., cattle shows or trade) [9]. In Belgian dairy herds, the absence of a dedicated calving pen and the use of a breeding bull significantly increased the risk of *M. bovis* presence in dairy herds [10]. Large herd size (greater than 100 animals), lack of quarantine procedures for newly purchased animals, and higher-than-average milk production were found to be major risk factors for *M. bovis* seropositivity in Brazilian dairy herds, managed under tropical conditions [11]. In a Japanese study, large dairy herds (greater than 200 animals) and herds that frequently purchased cattle were at higher risk of *M. bovis* infection [12]. While these studies provided seroprevalence estimates for lactating dairy cows, they did not include non-lactating animals such as replacement dairy heifers. Replacement dairy heifers constitute a key demographic in epidemiological investigations, as they can act as reservoirs for *M. bovis* and significantly influence intra-herd transmission dynamics [13]. The omission of non-lactating animals, such as heifers, from seroprevalence assessments limits the ability to thoroughly evaluate *M. bovis* transmission and its subsequent effects on herd health.

Risk factors associated with *M. bovis* infections in calves are varied and include several management practices and environmental conditions. One significant risk is the feeding of milk contaminated with *M. bovis*, such as waste milk from infected cows, which has been shown to facilitate transmission [14,15]. Additionally, the proximity of infected adult cows or calves to young calves increases the likelihood of infection. Infections frequently originate from cows with subclinical mastitis, which can shed the bacteria through milk or respiratory secretions, amplifying the risk of *M. bovis* transmission [16]. Calves housed in contaminated environments such as those with shared bedding or water sources, are also at heightened risk of exposure to *M. bovis*. Poor ventilation and overcrowding in housing facilities exacerbate this risk by contributing to the spread of respiratory infections [17]. Preweaning exposure to *M. bovis* in replacement dairy heifers significantly increases the likelihood of future pathogen shedding and disease transmission within the herd, highlighting the long-term impacts of early-life exposure [13].

Despite the reported morbidity and high herd prevalence of *M. bovis* in Ireland, the seroprevalence of this pathogen in dairy calves and youngstock remains largely unexplored. While previous studies have focused on adult cattle and lactating cows, there is a significant gap in understanding the infection dynamics among younger animals, which play a crucial role in the epidemiology of *M. bovis*. This study addresses a significant gap in the current understanding of *M. bovis* transmission by examining the seroprevalence in replacement dairy heifers during the critical rearing period (0–2 years of age). Additionally, it aims to determine herd-level risk factors associated with seropositivity in these youngstock cohorts. By focusing on this under-researched demographic, this study provides valuable insights into the spread of *M. bovis* and informs more effective control strategies for managing the infection within Irish dairy herds.

2. Materials and Methods

2.1. Herd Recruitment

In total, 120 dairy herds were recruited to a wider national longitudinal study to assess the animal health risks associated with contract-rearing in Ireland between 2018 and 2021 as described in McCarthy et al. [18]. Briefly, a national register of source dairy farms (SDF), (farmers who sent their heifers out to be contract-reared) and control farms (CF), (farmers who did not send their heifers out to be contract-reared) was generated. The 256 SDF were identified from the Irish Cattle Breeding Federation (ICBF) HerdPlus database. Control farms were matched to the SDFs by herd size, calving pattern, and geographical location. Telecontact with these farmers resulted in a database of 120 SDFs and 85 CFs. A total of 66 SDFs and 54 CFs were recruited from the database to participate in a longitudinal study of the animal health and production implications of contract-rearing [18–22]. The recruited farms were distributed across all 4 provinces and 19 of the 26 counties of the Republic of Ireland, with the largest density of farms located in County Cork, reflecting the distribution of the national dairy cow population. The majority of recruited herds were classified as spring-calving (92%), with the remaining herds operating a split-calving pattern (spring and autumn).

2.2. Farm Visit Schedule

A single cohort of heifer calves born in spring 2018 was followed until the end of their first lactation in 2021. Between spring 2018 and autumn 2019, SDFs and CFs were visited four times. The first visit occurred when heifers were approximately 1 month old, conducted on their farm of origin. Subsequent visits for home-reared heifers remained at the farm of origin, while for contract-reared heifers, visits were subsequently carried out on the contract-rearing unit. The second visit, when heifers were around 8 months old, occurred between September and December 2018. The third and fourth visits occurred when the heifers were approximately 12 months old in spring 2019 and 20 months old in autumn 2019. Approximately 6500 heifers were enrolled initially, with data available for 5532 heifers across all four visits after losses due to sales, farm drop-out, and mortality.

2.3. Blood Sampling

During the first (spring 2018), third (spring 2019), and fourth (autumn 2019) farm visits, blood samples were taken from 10 heifers on each farm. Heifers were sampled at random, typically with one heifer selected each time the handling facility (chute) was refilled, depending on its capacity and the size of the heifer cohort. Blood samples were not obtained at the second farm visit (autumn 2018) as the collection of other sample types, including faecal samples was prioritised at this visit.

Samples were initially obtained from young calves (<4 months old) via jugular venepuncture. At subsequent farm visits, blood samples were collected by coccygeal venepuncture from older animals. Samples were collected in plain vacutainer tubes (BD Vacutainer, BD, Langen, Germany) and stored in a cooler unit until returned to the research centre (Teagasc, Moorepark, Cork, Ireland). Samples were refrigerated for approximately 24 h post-collection, after which time serum was separated by centrifugation at $3500\times g$ for 15 min at 4 °C and frozen at −20 °C pending analysis at the conclusion of all farm visits.

Serum samples were analysed by a commercially accredited laboratory (FarmLab Diagnostics, Roscommon, Ireland) using the *M. bovis* ID Screen® *Mycoplasma bovis* antibody ELISA (IDVet, Montpellier, France). All analyses were performed according to the manufacturer's instructions. The ID Screen® *Mycoplasma bovis* antibody ELISA has a reported sensitivity of 95.7% and a specificity of 100%. The cut-off for a positive sample was an S/N (sample to negative control ratio) value of <0.6. Inconclusive test results were classified as negative in the data analysis. There is no *M. bovis* vaccine available in Ireland. The median age of sampled heifers was 1.5, 12.8, and 20.2 months during the first, second, and final farm visits, respectively. Details on the health status of the studied heifer population, including the detection rate of *M. bovis* in nasal swabs from calves with respiratory disease symptoms, have been previously reported by McCarthy et al. (2021) [18].

2.4. Biosecurity Survey

To assess the biosecurity status of study farms (and thus risk factors associated with *M. bovis*), all participating farmers were invited to complete a questionnaire relating to biosecurity and management practices on their farms in September 2018. The questionnaire was compiled as described in McCarthy et al. [20]. Briefly, the questionnaire was compiled following a systematic literature review to identify existing questionnaires with a significant biosecurity component. Approximately 30 questionnaires were assessed for suitability, but none met the study's aims. Instead, relevant questions were compiled from web-based herd health management tools and selected published surveys. Additional questions were added following consultation with a biosecurity expert group.

The survey covered various aspects of farm management, including herd characteristics, bioexclusion practices, calving and newborn calf management, unweaned heifer management, weaned heifer management, and herd vaccination protocols (outlined in Table 1). The survey consisted of a combination of open- and closed-ended, multiple choice, and Likert scale questions. It was piloted and modified based on feedback to ensure completion within 30 to 40 min. Data processing involved inspecting responses, validating information with national databases, and coding responses for analysis. Descriptive statistics were calculated for farm characteristics and management practices.

Table 1. Farm management and associated variables included in questionnaires administered to participating dairy farmers.

Questionnaire Section	Variables
Farmer and herd characteristics	Address, herd number, herd designator, reason for participation, calving pattern, herd size, replacement rate, farm enterprise and other stock kept on the farm, herd biosecurity status, number of land blocks managed.
Bioexclusion practices	Use of own equipment/contractor to spread slurry, use of slurry from other herds, grazing management after slurry application. Use of footbaths, livestock transportation method, cleaning/disinfection practices for farm visitors, co-grazing of animals with other species, hygiene practices when working with animals of different ages, rodent control policy, and farm water source. Number of animals purchased and source of bought-in animals, isolation and testing policy for bought-in animals. Details of bounding farms and rivers/streams running through farms. Pets are kept on farms, and wildlife is seen on farms.
Calving and newborn calf management	Type of calving facility used, max number of animals it can house, when cows are moved to calving pen, location of calving facilities, use of calving facilities for sick animals, cleaning and disinfection procedures of calving facilities. Navel disinfection practices, product, and timing of application. Colostrum feeding practices; source, quantity, and method of feeding; assessment of quality and storage.
Unweaned heifer management	Type of milk fed to calves, number of feedings per day, feeding of nonsaleable milk, calf housing, sick calf housing, calf feeding equipment, cleaning/disinfection of housing facilities, weaning criteria.
Weaned heifer management	Grazing management and parasite control strategies.
Herd vaccination protocol (biocontainment practices)	Products used and dates of vaccination for the following: BVD [1], IBR [2], calf pneumonia, calf diarrhoea, leptospirosis, salmonellosis.
Specific questions for source dairy farmers	Age/weight/month heifers moved to CR responsibility for breeding management, health checks performed on heifers before moving, transport of heifers to and from CR, isolation of heifers upon return from CR.

[1] BVD: Bovine viral diarrhoea; [2] IBR: Infectious bovine rhinotracheitis.

2.5. Statistical Analysis

2.5.1. Data Cleaning

This study employed a repeated cross-sectional design to assess herd-level seroprevalence of *M. bovis* across multiple farm visits. For each farm visit, the individual *M. bovis* ELISA test results from animals were aggregated at the herd level. Herds were classified as seropositive based on two criteria: (1) if at least one animal (out of a sample of ten) tested seropositive, and (2) if three or more animals (out of a sample of ten) tested seropositive. Therefore, the outcome variable for the analysis was binary, indicating whether the heifer cohort was seropositive or seronegative for *M. bovis*. Two models were then developed for each farm visit: Model $\geq$ 1POS, where a herd was considered seropositive if at least one heifer tested seropositive, and Model $\geq$ 3POS, where a herd was considered seropositive if at least three heifers tested seropositive.

The independent variables (n = 59) of interest were biosecurity and herd management factors gathered through the biosecurity questionnaire survey. Where possible, survey questions with multiple response options were dichotomised. All ELISA results and survey data were compiled in a Microsoft Excel spreadsheet and exported to R studio (Version 2024.04.2) for further analysis. Herds for which ELISA results were not available at all three visit periods were removed from the dataset, resulting in a final dataset of 105 herds available for analysis.

2.5.2. Data Analysis

Multivariable logistic regression analysis was utilised to quantify the associations between various potential risk factors and *M. bovis* serostatus. This approach allowed for the estimation of the effect of each predictor variable while adjusting for potential confounders and covariates.

Multivariable Logistic Regression Models

Descriptive statistics were conducted using R studio (Version 2024.04.2). Continuous variables were examined visually for normality using histogram plots. Non-normally distributed continuous variables were log-transformed.

To identify herd-level risk factors associated with *M. bovis* seropositivity, three multivariable logistic regression models (corresponding to the three sampling periods) were built for each of the two criteria ($\geq$1 positive animal and $\geq$3 positive animals); a total of six models. The outcomes of interest in the models were seropositivity in spring 1, seropositivity in spring 2, and seropositivity in autumn 2. All analyses were performed at the herd level. By concentrating on herd-level risk factors, this study aimed to provide actionable recommendations that could be implemented at the farm-level, potentially offering more effective and scalable solutions for managing *M. bovis* compared to focusing on individual animal-level factors. Potential risk factors were first screened for association with *M. bovis* serostatus at a univariable level. Variables with a p value of approximately 0.2 were brought forward to the multivariable analysis. Pearson correlation coefficients were used to assess the correlation between predictor variables. Pairs of highly correlated variables were identified using a correlation matrix (i.e., with a correlation coefficient of >0.8). To avoid bias and overfitting, the final variable selection was informed by biological relevance and causal relationships. Variables with strong biological or causal significance were prioritised. Missing data were removed before building the initial multivariable model resulting in a reduction in the number of herds available for the final multivariable model (see Table 7 for details of the number of herds for which a complete dataset was available for each visit period).

Multivariable forward stepwise models were then constructed for the three sampling periods of interest. Variables were added to the model based on their significance until adding more variables did not significantly improve the model fit (based on reduction of AIC). Backward stepwise selection was then conducted, resulting in the inclusion of the same variables, indicating robustness in the selection process.

Potential two-way interactions between variables were evaluated by incorporating interaction terms into the logistic regression model, and their significance was assessed using p values, with a threshold of $\leq$0.05 considered indicative of a statistically significant interaction. There were no biologically meaningful interactions evident. A significance level of 5% was used in all models.

To evaluate the performance of the models, McFadden's R-squared was used. This metric assesses how much better the fitted model is compared to a null model (i.e., a model without predictors), which serves as a reference point for comparison. Higher McFadden's

R-squared values indicate a better model fit, suggesting that the model more effectively accounts for the variability in *M. bovis* seropositivity.

3. Results

3.1. Descriptive Statistics

In total, *M. bovis* ELISA results were available for 105 herds across all three farm visit periods. The median herd size was 141 cows (range 60–633 cows) with a median heifer cohort size of 41 heifers (range 10–137 heifers). The median age of heifers at each visit period is outlined in Table 2. The median age of the randomly selected sampled heifers was comparable to that of their herd mates within each herd at each visit, ensuring that the sampled animals were representative of the entire heifer cohort on the farm.

Table 2. Median age (in days) of heifer cohort and sampled heifers at each sampling period (n = 105 herds).

Visit Period	Heifers (n)	Heifer Cohort Age (days)	Sampled Heifer Cohort Age (days)
V1 (Spring 2018)	1065	41 (range 1–112)	45 (1–103)
V2 (Spring 2019)	1091	387 (range 265–456)	388 (304–456)
V3 (Autumn 2019)	1090	613 (range 481–694)	613 (483–691)

Whenever possible, 10 heifers were sampled per herd during each visit period. However, there was a slight variation in the number of samples collected at each visit due to fluctuations in the heifer cohort size at the time of sampling. This variation was particularly evident during the first visit period in spring 2018 when a proportion of the heifer cohort had not yet been born. Despite these fluctuations, 10 samples were successfully collected in 89%, 88%, and 92% of farms during visit periods 1, 2, and 3, respectively. The distribution of herds with at least one *M. bovis* seropositive heifer and three of more seropositive heifers is shown in Table 3.

Table 3. Distribution of herds with at least one *M. bovis* seropositive heifer and at least 3 seropositive heifers across three farm visit periods between spring 2018 and autumn 2019 (n = 105 herds).

	Seropositivity Criteria	
	≥1 Positive heifer	≥3 Positive heifers
Visit period	Herd seroprevalence (%) (95% CI)	Herd seroprevalence (%) (95% CI)
V1 (Spring 2018)	50.4 (40.5–60.3)	31.4 (22.7–41.2)
V2 (Spring 2019)	35.2 (26.2–45.1)	32.4 (23.6–42.2)
V3 (Autumn 2019)	45.7 (36.0–55.7)	42.9 (33.2–52.9)

In total, approximately 3200 samples were collected across all three farm visit periods. The distribution of sampling frequency at the individual heifer level was as follows; 69% of the heifers were sampled once (n = 2218), 13.4% were sampled twice (n = 432) and 1.7% were sampled three times (n = 54). The mean *M. bovis* apparent seroprevalence (within the positive herds) during each visit period is outlined in Table 4.

Table 4. Mean (se) within-herd *M. bovis* apparent seroprevalence and standard error across three visit periods (10 heifers sampled per herd).

Visit Period	Herds (n = 48 CF, 57 SDF *)	Herds with at Least One *M. bovis*-Positive Heifer (n)	Mean (95% CI) *M. bovis* Seropositivity (%)
V1 (Spring 2018)	105	53	38.3 (31.5–45.1)
V2 (Spring 2019)	105	37	81.3 (72.1–90.5)
V3 (Autumn 2019)	105	48	77.1 (69.0–85.2)

* CF: Control farm: rearing their own heifers, SDF: Source dairy farm: engage in contract-rearing of heifers.

The distribution of farms by *M. bovis* seropositivity percentage by farm visit period is demonstrated in Figure 1.

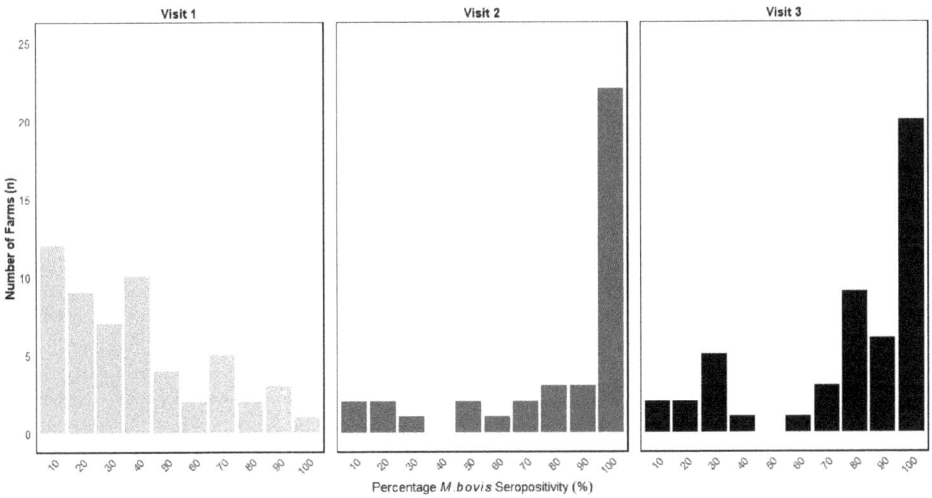

Figure 1. Distribution of herds (n = 105) by percentage *M. bovis* seropositivity across three farm visit periods between spring 2018 and autumn 2019.

Animal-level apparent seroprevalence is demonstrated in Table 5.

Table 5. Animal-level seroprevalence of *M. bovis* in heifers over three farm visit periods. The table shows the number of heifers sampled, the number testing positive for *M. bovis* antibodies, and the corresponding seropositivity percentages for spring 2018 (V1), spring 2019 (V3), and autumn 2019 (V3) (n = 105 herds).

Visit Period	Heifers (n)	Positive *M. bovis* ELISA (n)	*M. bovis* Seropositivity (%) (95% CI)
V1 (Spring 2018)	1065	214	20.1 (17.7–22.5)
V2 (Spring 2019)	1091	319	29.2 (26.5–31.9)
V3 (Autumn 2019)	1090	393	36.1 (33.2–38.9)

The variation in the herd *M. bovis* seroprevalence across the three visit periods using the two positive threshold values of (≥1 positive heifer; ≥3 positive heifers) is outlined in Table 6.

Table 6. Number of herds classified as *M. bovis* seropositive or seronegative by two seropositivity criteria, across three farm visit periods (n = 105).

Herd *M. bovis* Serostatus Category			Herds (n)	
Visit 1	Visit 2	Visit 3	≥1 Positive Heifer	≥3 Positive Heifers
Positive	Positive	Negative	0	0
Negative	Positive	Negative	0	0
Positive	Negative	Positive	3	1
Negative	Negative	Positive	8	10
Positive	Negative	Negative	23	14
Negative	Positive	Positive	10	16
Negative	Negative	Negative	34	46
Positive	Positive	Positive	27	18

Key: visit 1 (spring 2018), visit 2 (spring 2018), visit 3 (autumn 2019).

3.2. Univariable Models

Univariable analysis was conducted initially to investigate the association between biosecurity and farm management practices and herd-level *M. bovis* seropositivity at each of the three visit periods (Table S1, Supplementary Materials).

In spring 1 (V1), in Model $\geq$ 1POS, the housing of heifer and bull calves separately had a borderline association with seropositivity (estimate: 0.78, p = 0.06). A greater number of colostrum feeds was associated with a slight decrease in seropositivity (estimate: -0.19, p = 0.05). Herd size was marginally positively associated with seropositivity, suggesting that larger herds were more likely to be seropositive (estimate: 0.003, p = 0.05). Additionally, the presence of non-dairy animals on the farm was marginally associated with increased seropositivity (estimate: 0.89, p = 0.07).

In spring 2 (V2), in Model $\geq$ 1POS, the practice of evaluating colostrum quality showed a non-significant trend towards higher seropositivity (estimate: 0.65, p = 0.12), while feeding waste milk to calves also indicated a potential increase in risk (estimate: 1.07, p = 0.07). The purchase of cattle during 2018 was a significant risk factor (estimate: 1.00, p = 0.05). Similar to Model $\geq$ 1POS, cattle purchase during 2018 (estimate: 0.83, p = 0.11) and farming multiple land parcels (estimate: 0.92, p = 0.04) were associated with increased risk of seropositivity in Model $\geq$ 3POS. Feeding waste milk to calves again showed a non-significant association with increased seropositivity (estimate: 0.92, p = 0.12).

During autumn 2 (V3), in Model $\geq$ 1POS, herds in which farmers assessed colostrum quality were more likely to be seropositive, with an estimate of 0.79 (p = 0.05). Similarly, the purchase of cattle in 2018 was significantly associated with seropositivity (estimate: 0.95, p = 0.04).

Several associations did not reach conventional levels of statistical significance ($p < 0.05$) at a univariable level, which may reflect limitations in this study's statistical power due to the sample size. This potential lack of power could explain why some trends observed in the analysis did not achieve significance, despite suggestive estimates.

3.3. Multivariable Analysis

Variables with a p value of approximately 0.2 were selected for inclusion in the initial multivariable models for each visit period using the two herd seropositivity thresholds to determine herd *M. bovis* status. The results of the final multivariable models are presented in Table 7.

3.4. Model Performance

The comparison of McFadden's R-squared values between Model $\geq$ 1POS and Model $\geq$ 3POS across the three visit periods demonstrates variations in model performance over time (Table 8). In visit 1 (spring 1), Model $\geq$ 1POS had an R-squared value of 0.11, while Model $\geq$ 3POS had a similar value of 0.10. During visit 2 (spring 2), Model $\geq$ 3POS showed a stronger fit, with an R-squared value of 0.31 compared to 0.13 for Model $\geq$ 1POS. By visit 3 (autumn 2), both models performed equally, each achieving an R-squared value of 0.16. This indicates that Model $\geq$ 3POS had the best explanatory power in Spring 2, while both models showed similar predictive performance in Visits 1 and 3.

Table 7. Output from multivariable analysis of variables associated with heifer cohort *M. bovis* seropositivity across three farm visit periods (spring 1, spring 2, and autumn 2) in two models with different criteria for heifer cohort seropositivity. These variables represent the final selection after univariable analysis.

	Model ≥ 1POS: One or More ELISA Positive Heifers Indicate Herd Seropositivity					**Model ≥ 3POS: Three or More ELISA Positive Heifers Indicate Herd Seropositivity**				
Visit Period	Variable Name	Estimate	SE	Odds ratio (95% CI)	*p* value	Variable Name	Estimate	SE	Odds ratio (95% CI)	*p* value
	(n = 99 herds, 46P, 53N, 46.46% seropositivity)					(n = 99 herds, 68N, 31P, 31.3% seropositivity)				
V1: Spring 1	Number of feeds of colostrum before whole milk or milk	−0.22	0.11	0.81 (0.64, 0.99)	0.05	Other animals kept on the farm (Reference category: No)	1.42	0.62	4.13 (1.24, 14.56)	0.02
	Pre-weaning individual housing only (Reference category: No)	0.85	0.53	2.33 (0.84, 6.93)	0.11	Pre-weaning individual housing only (Reference category: No)	1.09	0.60	2.98 (0.92, 10.06)	0.07
	Heifer calves are housed separately from bull calves (Reference category: No)	0.95	0.47	2.60 (1.05, 6.78)	0.04	Herd size in 2018	0.005	0.00	1.0 (1.0, 1.01)	0.12
	Navel disinfection carried out (Reference category: No)	1.18	0.77	3.25 (0.79, 17.18)	0.12					
	Number of parcels of land farmed during 2018 (Reference category: 3 or less)	0.74	0.49	2.10 (0.82, 5.60)	0.13	Calves share airspace with older animals (Yes/No)	0.76	0.49	2.13 (0.82, 5.76)	0.12
	(n = 101 herds, 66N, 35P, 34.65% seropositivity)					(n = 99 herds, 68N, 31P, 31.3% seropositivity)				
V2: Spring 2	Feed waste milk to calves (Reference category: No)	1.08	0.65	2.93 (0.89, 1.84)	0.10	Feed waste milk to calves (Reference category: No)	1.04	0.68	2.84 (0.81, 12.33)	0.13
	Number of parcels of land farmed during 2018 (Reference category: 3 or less)	0.66	0.51	1.93 (0.71, 5.37)	0.20	Number of parcels of land farmed during 2018 (Reference category: 3 or less)	1.28	0.52	3.59 (1.33, 10.22)	0.01
	Purchased cattle during 2018 (Reference category: No)	1.35	0.60	3.84 (1.27, 13.78)	0.02	Purchased cattle during 2018 (Reference category: No)	1.22	0.62	3.40 (1.09, 12.60)	0.05
	Use group calving pens only (Reference category: No)	0.75	0.48	2.12 (0.84, 5.50)	0.12	Number of feeds of colostrum before whole milk or milk	−0.28	0.13	0.76 (0.57, 0.96)	0.04
	Implemented a vaccination protocol for IBR in calves during 2018 (Reference category: No)	0.77	0.47	2.16 (0.88, 5.48)	0.10					

Table 7. *Cont.*

Visit Period	Model ≥ 1POS: One or More ELISA Positive Heifers Indicate Herd Seropositivity					Model ≥ 3POS: Three or More ELISA Positive Heifers Indicate Herd Seropositivity				
	Variable Name	Estimate	SE	Odds ratio (95% CI)	*p* value	Variable Name	Estimate	SE	Odds ratio (95% CI)	*p* value
	(n = 103 herds, 60N, 43P, 46% seropositivity)					(n = 100, 60N, 40P, 40% seropositivity)				
Autumn 2	Calves share airspace with older animals (Reference category: No)	0.84	0.47	2.31 (0.94, 5.97)	0.07	Calves share airspace with older animals (Reference category: No)	0.94	0.49	2.56 (1.0, 1.23)	0.05
	Purchased cattle during 2018 (Reference category: No)	1.22	0.56	3.40 (1.20, 10.86)	0.03	Purchased cattle during 2018 (Reference category: No)	1.31	0.56	3.69 (1.30, 11.92)	0.02
	Heifer calves are housed separately from bull calves (Reference category: No)	1.30	0.50	3.66 (1.43, 10.20)	0.0092	Heifer calves are housed separately from bull calves (Reference category: No)	1.22	0.50	3.40 (1.31, 9.50)	0.01
	Colostrum quality assessed (Reference category: No)	0.90	0.46	2.44 (1.01, 6.13)	0.05	Colostrum from own dam only (Reference category: No)	−0.97	0.63	0.38 (0.10, 1.24)	0.12
	Feed waste milk to calves (Reference category: No)	0.83	0.60	2.30 (0.74, 7.95)	0.16	Number of parcels of land farmed during 2018 (Reference category: 3 or less)	1.14	0.51	3.13 (1.17, 8.91)	0.03
						Have a dedicated sick pen for calves (Reference category: No)	0.73	0.52	2.08 (0.77, 6.02)	0.16

Key: V1: visit 1 (spring 2018), V2: visit 2 (spring 2018), V3: visit 3 (autumn 2019).

Table 8. McFadden R-squared values comparing multivariable model performance across visit periods. Model ≥ 1POS indicates herd seropositivity based on one or more positive heifers, while Model ≥ 3POS is based on three or more positive heifers. Higher R-squared values reflect better model fit, showing the ability of the models to explain variance in *M. bovis* seropositivity across different visit periods.

Visit Period	Model ≥ 1POS	Model ≥ 3POS
Visit 1 (Spring 1)	0.11	0.10
Visit 2 (Spring 2)	0.13	0.31
Visit 3 (Autumn 2)	0.16	0.16

4. Discussion

4.1. Overview

The key findings from this study demonstrated several critical insights into the seroprevalence of *M. bovis* in heifers across three distinct time points during the rearing period. This is the first Irish study focused specifically on youngstock *M. bovis* seroprevalence, offering a unique perspective that contrasts with previous studies, which have primarily focused on the adult (lactating) cow cohort [8]. Internationally, similar studies have highlighted *M. bovis* seroprevalence in confined systems [23], but this research expands the understanding of pathogen dynamics within pasture-based, predominantly spring-calving dairy enterprises. This study highlights the dynamic nature of *M. bovis* infection in cattle herds, with seroprevalence rising over time as heifers aged and were repeatedly exposed, suggesting that ongoing transmission within the herd plays a significant role in maintaining infection, which may complicate disease control efforts.

4.2. Herd- and Animal-Level M. bovis Seroprevalence over Time

In spring 2018, 50.4% of herds had at least one seropositive heifer but only 31.4% had three or more seropositive animals. These findings suggest that while *M. bovis* was present in these herds, the pathogen had not yet been extensively transmitted between heifers in the cohort. The presence of seropositive animals could be attributed either to natural infections occurring within the calf cohort or to the detection of maternal-derived antibodies (MDA) in these young animals. MDAs, transferred through colostrum, can provide temporary immunity to calves, protecting them from early infections or reducing the severity of infection, which may contribute to detectable antibody levels even in the absence of active infection. Therefore, the observed seropositivity might reflect a combination of early, limited infections and the transient presence of these maternal antibodies in the young heifers.

By spring 2019 (V2), the proportion of herds with at least one *M. bovis*-positive heifer decreased to 35.2%. While this could potentially reflect the successful implementation of control measures in some herds, it may also be attributed to statistical variation, given the small number of animals sampled from larger populations in each herd, which could influence the observed changes in seroprevalence. However, the consistent proportion of herds with three or more seropositive heifers (32.4%) suggests that once *M. bovis* was established, it continued to spread within the herd. During this period, mean within-herd seropositivity sharply increased from 38.3% to 81.3%, potentially reflecting the heightened susceptibility of older heifers as their exposure to the pathogen increased. Increased exposure due to environmental and management practices likely contributed to this rise in seroprevalence. Studies have shown that *M. bovis* can persist in herds despite control efforts, often spreading more readily within herds that have not maintained stringent biosecurity measures [24,25].

The increase in seroprevalence at the animal- level, from 20.1% in spring 2018 to 29.2% in spring 2019, aligns with the understanding that older heifers, now lacking the protective effect of maternal antibodies, became more vulnerable to *M. bovis* infections. As a result, the pathogen spread more widely within those herds that had not fully contained the infection, leading to a higher overall infection rate.

In autumn 2019 (V3), the increase in the proportion of herds with at least one seropositive heifer to 45.7% suggests a resurgence of *M. bovis* infections. This could be due to the introduction of new, susceptible animals or seasonal factors that enhanced transmission, such as changes in housing or weather conditions. The rise in herds with three or more seropositive heifers to 42.9% reflects more extensive within-herd transmission, which is often challenging to control once established. Despite a slight decrease in mean within-herd seropositivity to 77.1%, the persistently high levels indicate that *M. bovis* continued to circulate widely within the affected herds. The animal-level seroprevalence, which reached 36.1%, underscores the cumulative impact of ongoing exposure and the pathogen's persistence over time.

This pattern is consistent with findings from other studies that highlight the difficulty in controlling *M. bovis* once it becomes endemic in a herd. Infections can persist and spread due to various factors, including the introduction of new animals, environmental stressors, and inadequate biosecurity measures. For instance, studies have shown that *M. bovis* can remain latent in some animals, re-emerging under stress or immunosuppression, which complicates eradication efforts [24,26].

Overall, the findings demonstrate that *M. bovis* seroprevalence increased both within herds and among individual animals as the heifers aged. The consistent increase in the number of heavily infected herds (with three or more positive animals) highlights the persistent challenge of controlling *M. bovis* and the need for continuous and adaptive management strategies. These results emphasise the importance of considering both age-related immunity and the dynamics of within-herd transmission when interpreting seroprevalence data and developing control measures for *M. bovis* in cattle populations. Given these trends, further investigation is warranted to better understand the underlying factors driving increased *M. bovis* transmission and to refine intervention strategies.

4.3. Risk Factors Associated with M. bovis Seropositivity in Replacement Heifers

This study utilised two distinct models to systematically evaluate risk factors associated with *M. bovis* seropositivity at three distinct time points during the heifer-rearing period. The results from both Model $\geq$ 1POS and Model $\geq$ 3POS provide critical insights into the epidemiology and management of *M. bovis* within dairy herds.

Number of land parcels.

An increased number of land parcels (four or more) farmed during 2018 was a significant risk factor for *M. bovis* seropositivity in both models across multiple visit periods. This association suggests that fragmented land management contributes to increased *M. bovis* transmission and infection rates within herds. These findings are consistent with those reported in an Irish study by McAloon et al. [8] who found that the number of contiguous farms was a risk factor for *M. bovis* bulk tank milk seropositivity in the lactating cow cohort. Each additional land parcel adds more shared boundaries and potential points of contact between cattle on adjoining farms, increasing the likelihood of interaction with multiple herds. This proximity can facilitate the spread of airborne and contact pathogens, especially when biosecurity measures are inconsistent across farms.

Land fragmentation is a prevalent characteristic of Irish dairy farms, with an average of six land parcels per farm, according to a study of 900 dairy farms [27]. Many of these farms are spread across multiple, non-contiguous parcels of land. In Bradfield's study, around 59% of farms managed 6 or fewer parcels, while 41% had between 7 and 22 distinct parcels. This fragmentation complicates farm operations by increasing the frequency of movement of animals, machinery/equipment (fomites), and workers between land parcels. This increased movement can facilitate the transmission of *M. bovis* between management groups. Several studies have shown that increased contact between different animal groups and the use of shared equipment without proper biosecurity measures can lead to higher risks of *M. bovis* transmission [4,10,23]. Additionally, managing multiple parcels of land increases the complexity of biosecurity protocols. Insufficient separation between infected and uninfected animals and the challenges of maintaining appropriate hygiene standards across different locations can further exacerbate the spread of *M. bovis* [28].

Purchase of cattle.

Purchase of cattle was a consistent factor for *M. bovis* seropositivity in both models, particularly in spring 2 and autumn 2. These findings agree with those of several studies which have demonstrated that purchased cattle act as a source of new *M. bovis* infections in dairy herds [10,12,29,30]. In a risk factor study by Burnens et al. [31], the purchase of animals was the only variable significantly associated with the *M. bovis* serological status of 51 dairy herds in Switzerland. Carrier cattle without clinical signs are the primary vectors for introducing *M. bovis*, often leading to varying transmission dynamics across herds. Some herds experience immediate clinical outbreaks, while others face delayed transmission events. The risk of *M. bovis* seropositivity increases significantly when animals are sourced from multiple herds, compared to single-source or closed herds [8]. Since the removal of EU milk quotas in 2015, Ireland's dairy sector has undergone considerable growth, with the national dairy herd expanding by approximately 45%, from 1.2 million cows in 2015 to nearly 1.6 million by 2021, accompanied by a 60% increase in milk production [32].

When herds expand through the purchase of animals, they introduce a significant biosecurity risk, as carrier cattle without clinical signs are a known vector for *M. bovis* transmission [31]. Of the herds enrolled in the current study, 93% had undergone expansion between 2013 and 2018, many by purchasing cattle, which likely increased their exposure to *M. bovis* compared to non-expanding, closed herds.

Therefore, the most effective way to prevent *M. bovis* infections is to maintain a closed herd [33]. Where this is not feasible, purchased animals should be tested and quarantined before introduction to the herd. Prior to the purchase of lactating cows, it is recommended that milk samples should be tested for *M. bovis* using culture, PCR, or ELISA. Additionally, calf health records should be reviewed for any history of *M. bovis*-related diseases, such as pneumonia or otitis media [24].

Other (non-dairy) animals are kept on the farm.

The presence of non-dairy animals (suckler beef cattle, finishing beef cattle, pigs, poultry, sheep, horses, or goats) on the farm was a significant risk factor for *M. bovis* seropositivity in replacement heifers in Model $\geq$ 3POS during the spring 1 visit period (V1). The odds ratio (OR 4.13) indicates that farms with non-dairy animals are substantially more likely to be *M. bovis* seropositive. While this finding has not been reported in other studies relating to *M. bovis*, a Norwegian study identified that the presence of "other animal traffic," including common animal housing and interaction between different animal types, was associated with higher risks of BVD in dairy herds [34]. Further to this, Correa-Valencia et al. [35] demonstrated that mixed farming enterprises had increased odds of infection with *Mycobacterium avium* subsp. *paratuberculosis* (MAP). A possible explanation for the increased seropositivity on mixed farm enterprises is that specialist dairy-only enterprises may maintain a higher level of control over biosecurity due to a singular focus on one species. These farms may have better implementation of preventative measures for dairy-specific diseases such as mastitis. Furthermore, dairy-only farms are less likely to experience interspecies disease transmission, reducing the overall disease burden on these farms.

Individual farm management practices.

Feeding waste milk (unpasteurised mastitic milk and milk from antibiotic-treated cows) to calves was identified as a potential risk factor for *M. bovis* seropositivity. While the association was not statistically significant, there was a consistent trend toward higher seropositivity in both spring 2 (V2) and autumn 2, indicating a potential but inconclusive association. This practice is linked to an elevated risk of transmitting several pathogenic agents, including *M. bovis* [9,14,15,24]. Hazelton et al. [13], and Gabinaitiene et al. [36] demonstrated that calves fed mastitic milk from cows infected with *M. bovis* are at a higher risk of becoming infected themselves. The association between feeding waste milk and higher *M. bovis* seroprevalence in the current study underscores the role of this practice in the spread of infection from the lactating cohort to the youngstock cohort within herds. It was surprising that feeding waste milk was not detected as associated with seropositivity in spring 1. This may reflect the time lag required between ingestion of a sufficient 'dose' of *M. bovis* infected colostrum/milk and expression of a detectable systemic antibody response. Many of the heifers at this visit period may not have been exposed to waste milk feeding practices for sufficiently long enough and may have been too young for detection of an antibody response (median age of the sampled heifers; 1.4 months old). Antibodies against *M. bovis* are not detected until 10–14 days post-infection but detectable levels of antibodies remain for several months to years [25,37,38], hence associations in spring 2 and autumn 2. Culling mastitic cows and either discarding or pasteurising infected colostrum and raw milk are, therefore, recommended to mitigate the risk of transmitting *M. bovis* to calves [39]. Pasteurisation of colostrum or waste milk is not commonly carried out on Irish dairy farms.

Shared airspace between calves and older animals was identified as a potential risk factor for *M. bovis* seropositivity, with a trend toward increased seropositivity in the studied herds. Although not always statistically significant, there was a consistent association across different visit periods and in both models. These findings are consistent with those reported by Gille et al. [10], who demonstrated that co-mingling of animals of different ages or groups, particularly in environments with suboptimal biosecurity, was a risk factor for increasing *M. bovis* transmission.

Older animals, particularly those that are sub-clinically infected or have recovered from clinical *M. bovis* infections, can serve as reservoirs of infection. *M. bovis* is primarily transmitted through respiratory secretions and close contact, making shared airspace, particularly in confined environments such as sheds, a critical risk factor for *M. bovis* transmission [40]. On farms where airspace is shared between animals of different age groups, there may be further shared environmental spaces, such as feeding areas, water troughs, and bedding. In these common areas, the potential for contamination with *M. bovis* through respiratory secretions or nasal discharge is considerable, potentially resulting in a higher pathogen load in the environment and a resultant increased risk of transmitting the infection to calves. In Irish spring-calving dairy systems, cows typically calve indoors during the early spring months. After birth, calves are kept indoors for approximately eight to twelve weeks. Calves commonly share the same airspace as the cows or, in some cases, the previous year's calves (now yearlings). This shared housing arrangement can pose challenges for disease control, particularly respiratory infections, due to the proximity and potential exposure to pathogens circulating among the older animals. Proper ventilation and separation are critical to minimising these risks, but in many cases, practical farm layouts mean calves are housed in close proximity to older animals, potentially exacerbating disease transmission. To reduce *M. bovis* transmission, calves should be housed separately from older cattle. Adequate ventilation, reduced stocking densities, and physical separation between calf and adult housing, can significantly lower *M. bovis* transmission risk [23,31].

The number of colostrum feeds was found to have a protective effect against *M. bovis* seropositivity, with calves receiving more colostrum feeds showing lower odds of seropositivity across multiple visit periods. This aligns with existing research that highlights the critical role of colostrum in providing passive immunity to newborn calves. Colostrum contains high levels of immunoglobulins, which are essential for the early development of the immune system in calves [41]. Studies have consistently shown that inadequate or delayed colostrum intake significantly increases susceptibility to respiratory infections and systemic diseases such as *M. bovis* [42]. Thus, ensuring that calves receive an adequate volume of high-quality colostrum within the first hours of life is critical for enhancing immunity and reducing the risk of pathogen transmission in herds.

The practice of housing heifer calves separately from bull calves was significantly associated with *M. bovis* seroprevalence in both models. This may indicate a response by such farmers to reduce the risk of transmitting respiratory infections on their farms, some of which may be due to *M bovis* infection. However, as the initial source of *M bovis* infection in young calves is from colostrum/milk and the calving environment, subsequent separation of the heifer and bull calves may be less effective in reducing infection risk.

Other farm management practices.

Navel disinfection was associated with a tendency towards a higher likelihood of *M. bovis* seropositivity during spring 1. *M. bovis* may be transmitted indirectly between a cow and her environment and her calf shortly after birth when the farmer has calved the cow and then handled the calf to disinfect its navel. The open umbilicus is a conduit for pathogens causing omphalitis [43] but possibly also for other maternal or environmental pathogens, including *M. bovis*, introduced as contaminants. The positive association of navel disinfection with *M. bovis* seropositivity may also indicate broader hygiene issues on farms implementing this protocol.

The current study did not identify a significant association between contract heifer rearing and *M. bovis* seropositivity. One possible explanation for the lack of association observed in this study could be the variability in *M. bovis* infection risks across different contract-rearing facilities. Previous studies indicate that while certain facilities have higher infection rates, this risk is not consistent across all operations. Differences in management practices, biosecurity measures, and the presence of asymptomatic carriers could vary significantly between facilities, potentially masking any clear association within the study population [13].

4.4. Model Comparison

During visit 1 (spring 1), both Model $\geq$ 1POS and Model $\geq$ 3POS show relatively low R-squared values (0.11 and 0.10, respectively), indicating a modest ability to explain the variance in seropositivity. In visit 2 (spring 2), however, Model $\geq$ 3POS demonstrates a substantially better fit (R-squared = 0.31) compared to Model $\geq$ 1POS (R-squared = 0.13), suggesting that a stricter criterion of three or more positive heifers improves the model's ability to capture the dynamics of seropositivity during this period.

Conversely, in visit 3 (autumn 2), both models performed equally well, with identical R-squared values of 0.16, implying that the choice of threshold (one or three positive heifers) has no significant impact on model fit for this period. Overall, these results suggest that while the stricter criterion enhances model performance in certain contexts, particularly in visit 2, its advantage may be less pronounced in other periods.

Hughes et al. [44] demonstrated that McFadden's R-squared values for logistic regression are generally lower than R-squared values for linear models. Values between 0.2 and 0.4 are considered to indicate a good model fit in epidemiological models of disease risk. The McFadden R-squared values reported for Model $\geq$ 3POS, ranging from 0.16 to 0.31 across visit periods, suggest a reasonably good fit, particularly in visit 2 (spring 2), where the value reaches 0.31. This indicates that Model $\geq$ 3POS is better at capturing key infection risk factors, particularly in certain periods, compared to Model $\geq$ 1POS, which has consistently lower R-squared values (0.11 to 0.16).

The multivariable analysis reveals that both models consistently identified key risk factors, such as the number of land parcels farmed, the purchase of cattle, and the number of colostrum feeds as significant across all visit periods. This consistency highlights the importance of management practices and environmental exposures in influencing *M. bovis* transmission. Model $\geq$ 3POS tended to capture more severe infection risk factors and exhibited stronger predictive performance overall. Meanwhile, Model $\geq$ 1POS identified broader risk patterns across herds with varying levels of infection. The results suggest that stricter thresholds for defining herd seropositivity (as in Model $\geq$ 3POS) may be more effective in detecting higher-risk herds.

4.5. Study Limitations

This study has specific limitations that may constrain the broader applicability and interpretation of its results. One key limitation is the potential for selection bias in herd recruitment, as the farms were initially recruited for a separate study assessing the health impacts of contract-rearing of replacement heifers, though a substantial number of these were control herds. As a result, recruited herds may not fully represent the national dairy herd population. Furthermore, the voluntary nature of participation could have resulted in an over-representation of farmers with a greater interest in biosecurity and animal health, potentially skewing the results.

The initial sample size for the contract-rearing study was calculated based on the power requirements specifically needed to assess differences in age at first calving (AFC) between contract-reared and home-farm-reared heifers. Consequently, the sample size may not have been optimally designed to assess the seroprevalence of *M. bovis* at herd-level and it may have limited power for detecting significant associations between risk factors and *M. bovis* seropositivity. This could potentially affect the robustness and external validity of the findings related to *M. bovis* infection, highlighting a need for cautious interpretation and possibly further targeted studies with appropriate power calculations focused on *M. bovis* seroprevalence.

Additionally, a complete dataset was not available for every farm at each visit period (*M. bovis* ELISA results were available for 105 farms at all three farm visit periods; 120 farms were initially enrolled in this study). Incomplete datasets were predominantly a result of missing biosecurity and farm management variables, which limited the ability to comprehensively assess the associations between risk factors and *M. bovis* seropositivity across all farms, potentially weakening the statistical power of this study. Furthermore, the biosecurity and farm management data were collected through self-reported questionnaires, which carry the risk of biases, such as recency and social desirability bias, and may not reflect the actual practices on farms. In addition, biosecurity data were collected at one time point in 2018, which does not account for any subsequent changes in management practices during the study period.

This study followed a single cohort of heifers longitudinally from birth to their first lactation, but contemporaneous, unrecorded, variations in farm management, environmental conditions, and disease challenges could have influenced the results. The blood sampling was limited to 10 randomly selected heifers per herd, which may not fully represent the infection status of the entire herd or heifer cohort. Additionally, this study faced a considerable loss to follow-up, with data available for fewer heifers than initially enrolled, which may have introduced bias if the lost heifers differed in their exposure to risk factors compared to those that remained in this study. These factors collectively necessitate a cautious interpretation of the findings.

Despite the limitations of this study, it significantly advances our understanding of the epidemiology of *M. bovis* in dairy herds. This study identifies several herd-level risk factors, such as the number of land parcels farmed, purchase of cattle, and number of colostrum feeds as predictors of *M. bovis* seropositivity in replacement heifer cohorts. These findings are particularly important for guiding the development of targeted biosecurity measures aimed at controlling *M. bovis* transmission in youngstock. By identifying risk factors associated with *M. bovis* seroprevalence, this study contributes to ongoing efforts to mitigate the impact of this globally significant pathogen in the dairy industry.

5. Conclusions

In conclusion, this study provides valuable insights into the factors associated with *M. bovis* seropositivity in replacement dairy heifers at different time points during the rearing period. The findings consistently highlight key risk factors influencing *M. bovis* seropositivity, including the degree of land fragmentation, purchase of cattle, and specific management practices, such as feeding waste milk and shared airspace between calves and older animals. The results underline the importance of herd-level interventions, robust biosecurity protocols, and targeted management strategies in mitigating the spread of *M. bovis* within dairy operations. Future research should aim to address the gaps identified, further refining our understanding of risk factors and optimising control measures for this pathogen.

Supplementary Materials: The following supporting information can be downloaded at: https://www.mdpi.com/article/10.3390/ani14213057/s1, Table S1: Output from univariable analysis of variables associated with heifer cohort *M. bovis* seropositivity across three farm visit periods (spring 1, spring 2 and autumn 2) in two models with different criteria for cohort seropositivity (at a significance level of approximately $p < 0.2$). These variables were subsequently used to build the multivariable models.

Author Contributions: M.-C.M.: investigation, resources, data curation, formal analysis, writing of original draft, review, and editing. J.F.M.: conceptualisation, methodology, investigation, resources, project administration, funding acquisition, supervision, writing—review and editing. C.G.M.: supervision, methodology, writing—review and editing. L.O.: supervision, conceptualisation, methodology, writing—review and editing. All authors have read and agreed to the published version of the manuscript.

Funding: This research study was funded by the Teagasc Walsh Scholarship Fund (project MKAB-0146).

Institutional Review Board Statement: Ethical approval to conduct this study was obtained from the Teagasc Animal Ethics Committee (TAEC177-2017) and procedure authorization (AE19132/P075) was granted by the Health Products Regulatory Authority of Ireland (HPRA). The experiment was carried out in accordance with the European Union (Protection of Animals Used for Scientific Purposes) Regulations 2012 (S.I. No. 543 of 2012).

Informed Consent Statement: At the point of study recruitment, all farmers were informed about the purpose and scope of the study. They received a verbal briefing on what participation would entail, including data collection methods and the intended use of the data. At this stage, they provided verbal consent to participate in the study. Afterwards, all participants were provided with a consent form that granted permission for the research team to access all data related to their herds, including the health and performance data of individual animals. All study participants signed this form.

Data Availability Statement: The data used to support the findings of this study are included within the article, and the data are available from the corresponding author upon reasonable request.

Acknowledgments: The authors sincerely thank the farmers who participated in this study. We also thank the various farming stakeholder bodies who aided with farmer recruitment. In particular, we thank Jonathon Kenneally, Chloe Millar, Noel Byrne, and Kieran McCarthy for providing technical assistance for this work. The authors wish to thank Cosme Sánchez-Miguel (Cork RVL) for assistance with data analysis.

Conflicts of Interest: The authors declare no conflicts of interest.

Abbreviations

M. bovis—*Mycoplasma bovis*; CF—control farm; SDF—source dairy farm; ELISA—enzyme-linked immunosorbent assay; CI—confidence interval; OR—odds ratio; BVD—bovine viral diarrhoea; IBR—infectious bovine rhinotracheitis; BTM—bulk tank milk; ICBF—Irish Cattle Breeding Federation; PCR—polymerase chain reaction; MAP—*Mycobacterium avium* subsp. *paratuberculosis*; SE—standard error; HPRA—Health Products Regulatory Authority.

References

1. Maunsell, F.P.; Donovan, G.A. *Mycoplasma bovis* Infections in Young Calves. *Vet. Clin. N. Am. Food Anim. Pract.* **2009**, *25*, 139–177. [CrossRef] [PubMed]
2. Caswell, J.L.; Bateman, K.G.; Cai, H.Y.; Castillo-Alcala, F. *Mycoplasma bovis* in Respiratory Disease of Feedlot Cattle. *Bov. Respir. Dis.* **2010**, *26*, 365–379. [CrossRef] [PubMed]
3. Radaelli, E.; Castiglioni, V.; Losa, M.; Benedetti, V.; Piccinini, R.; Nicholas, R.; Scanziani, E.; Luini, M. Outbreak of Bovine Clinical Mastitis Caused by *Mycoplasma bovis* in a North Italian Herd. *Res. Vet. Sci.* **2011**, *91*, 251–253. [CrossRef] [PubMed]
4. Schibrowski, M.L.; Gibson, J.S.; Hay, K.E.; Mahony, T.J.; Barnes, T.S. *Mycoplasma bovis* and Bovine Respiratory Disease: A Risk Factor Study in Australian Feeder Cattle. *Prev. Vet. Med.* **2018**, *157*, 152–161. [CrossRef]
5. Rosengarten, R.; Citti, C. The Role of Ruminant Mycoplasmas in Systemic Infection. *Mycoplasmas Rumin. Pathog. Diagn. Epidemiol. Mol. Genet.* **1999**, *3*, 14–17.
6. Doherty, M.L.; McElroy, M.C.; Markey, B.K.; Carter, M.E.; Ball, H.J. Isolation of *Mycoplasma bovis* from a Calf Imported into the Republic of Ireland. *Vet. Rec.* **1994**, *135*, 259–260. [CrossRef]

7. DAFM All Island Disease Surveillance Report. Available online: http://www.animalhealthsurveillance.agriculture.gov.ie/media/animalhealthsurveillance/2022-All-Island-Animal-Disease-Surveillance.v.2.pdf (accessed on 10 September 2024).
8. McAloon, C.I.; McAloon, C.G.; Tratalos, J.; O'Grady, L.; McGrath, G.; Guelbenzu, M.; Graham, D.A.; O'Keeffe, K.; Barrett, D.J.; More, S.J. Seroprevalence of *Mycoplasma bovis* in Bulk Milk Samples in Irish Dairy Herds and Risk Factors Associated with Herd Seropositive Status. *J. Dairy Sci.* **2022**, *105*, 5410–5419. [CrossRef]
9. Aebi, M.; van den Borne, B.H.; Raemy, A.; Steiner, A.; Pilo, P.; Bodmer, M. *Mycoplasma bovis* Infections in Swiss Dairy Cattle: A Clinical Investigation. *Acta Vet. Scand.* **2015**, *57*, 10. [CrossRef]
10. Gille, L.; Callens, J.; Supré, K.; Boyen, F.; Haesebrouck, F.; Van Driessche, L.; van Leenen, K.; Deprez, P.; Pardon, B. Use of a Breeding Bull and Absence of a Calving Pen as Risk Factors for the Presence of *Mycoplasma bovis* in Dairy Herds. *J. Dairy Sci.* **2018**, *101*, 8284–8290. [CrossRef]
11. Pires, D.R.; Morais, A.C.N.; Cunha, N.C.; Machado, L.S.; Barbosa, L.F.C.; Mendonça, J.F.M.; Balaro, M.F.A.; Santos, J.P.C.; Souza, G.N.; Barreto, M.L. Proposal of an iELISA for *Mycoplasma bovis* Diagnosis in Dairy Cattle and Associated Risk Factors. *Arq. Bras. Med. Veterinária E Zootec.* **2021**, *73*, 293–301. [CrossRef]
12. Murai, K.; Higuchi, H. Prevalence and Risk Factors of *Mycoplasma bovis* Infection in Dairy Farms in Northern Japan. *Res. Vet. Sci.* **2019**, *123*, 29–31. [CrossRef] [PubMed]
13. Hazelton, M.S.; Morton, J.M.; Parker, A.M.; Bosward, K.L.; Sheehy, P.A.; Dwyer, C.J.; Niven, P.G.; House, J.K. *Mycoplasma bovis* and Other Mollicutes in Replacement Dairy Heifers From *Mycoplasma bovis*-Infected and Uninfected Herds: A 2-Year Longitudinal Study. *J. Dairy Sci.* **2020**, *103*, 11844–11856. [CrossRef] [PubMed]
14. Butler, J.A.; Sickles, S.A.; Johanns, C.J.; Rosenbusch, R.F. Pasteurization of Discard Mycoplasma Mastitic Milk Used to Feed Calves: Thermal Effects on Various Mycoplasma. *J. Dairy Sci.* **2000**, *83*, 2285–2288. [CrossRef] [PubMed]
15. Walz, P.H.; Mullaney, T.P.; Render, J.A.; Walker, R.D.; Mosser, T.; Baker, J.C. Otitis Media in Preweaned Holstein Dairy Calves in Michigan Due to *Mycoplasma bovis*. *J. Vet. Diagn. Investig.* **1997**, *9*, 250–254. [CrossRef]
16. Giovannini, S.; Zanoni, M.G.; Salogni, C.; Cinotti, S.; Alborali, G.L. *Mycoplasma bovis* Infection in Respiratory Disease of Dairy Calves Less than One Month Old. *Res. Vet. Sci.* **2013**, *95*, 576–579. [CrossRef]
17. Tschopp, R.; Bonnemain, P.; Nicolet, J.; Burnens, A. Epidemiological Study of Risk Factors for *Mycoplasma bovis* Infections in Fattening Calves. *Schweiz. Arch. Tierheilkd.* **2001**, *143*, 461–467.
18. McCarthy, M.-C.; O'Grady, L.; McAloon, C.G.; Mee, J.F. The Effect of Contract-Rearing on the Health Status of Replacement Dairy Heifers. *Animals* **2021**, *11*, 3447. [CrossRef]
19. McCarthy, M.-C.; O'Grady, L.; McAloon, C.G.; Mee, J.F. Longitudinal Prevalence of Antibodies to Endemic Pathogens in Bulk Tank Milk Samples from Dairy Herds Engaged or Not in Contract Heifer Rearing. *Front. Vet. Sci.* **2021**, *8*, 785128. [CrossRef]
20. McCarthy, M.C.; O'Grady, L.; McAloon, C.G.; Mee, J.F. A Survey of Biosecurity and Health Management Practices on Irish Dairy Farms Engaged in Contract-Rearing. *J. Dairy Sci.* **2021**, *104*, 12859–12870. [CrossRef]
21. McCarthy, M.-C.; Mee, J.F.; McAloon, C.G.; O'Grady, L. A Comparison of the Age at First Calving of Contract-Reared versus Home-Reared Replacement Dairy Heifers. *Theriogenology* **2022**, *181*, 105–112. [CrossRef]
22. McCarthy, M.-C.; McAloon, C.G.; O'Grady, L.; Mee, J.F. Growth Rates of Contract-Reared versus Home-Reared Replacement Dairy Heifers. *Animal* **2022**, *16*, 100570. [CrossRef] [PubMed]
23. Haapala, V.; Vähänikkilä, N.; Kulkas, L.; Tuunainen, E.; Pohjanvirta, T.; Autio, T.; Pelkonen, S.; Soveri, T.; Simojoki, H. *Mycoplasma bovis* Infection in Dairy Herds—Risk Factors and Effect of Control Measures. *J. Dairy Sci.* **2021**, *104*, 2254–2265. [CrossRef] [PubMed]
24. Maunsell, F.P.; Woolums, A.; Francoz, D.; Rosenbusch, R.; Step, D.; Wilson, D.J.; Janzen, E. *Mycoplasma bovis* Infections in Cattle. *J. Vet. Intern. Med.* **2011**, *25*, 772–783. [CrossRef] [PubMed]
25. Vähänikkilä, N.; Pohjanvirta, T.; Haapala, V.; Simojoki, H.; Soveri, T.; Browning, G.F.; Pelkonen, S.; Wawegama, N.K.; Autio, T. Characterisation of the Course of *Mycoplasma bovis* Infection in Naturally Infected Dairy Herds. *Vet. Microbiol.* **2019**, *231*, 107–115. [CrossRef]
26. Nicholas, R.A.J.; Ayling, R.D. *Mycoplasma bovis*: Disease, Diagnosis, and Control. *Res. Vet. Sci.* **2003**, *74*, 105–112. [CrossRef]
27. Bradfield, T.; Butler, R.; Dillon, E.; Hennessy, T.; Kilgarriff, P. The Effect of Land Fragmentation on the Technical Inefficiency of Dairy Farms. *J. Agric. Econ.* **2021**, *72*, 486–499. [CrossRef]
28. Arede, M.; Nielsen, P.K.; Ahmed, S.S.U.; Halasa, T.; Nielsen, L.R.; Toft, N. A Space–Time Analysis of *Mycoplasma bovis*: Bulk Tank Milk Antibody Screening Results from All Danish Dairy Herds in 2013–2014. *Acta Vet. Scand.* **2016**, *58*, 16. [CrossRef]
29. Punyapornwithaya, V.; Fox, L.K.; Hancock, D.D.; Gay, J.M.; Alldredge, J.R. Association between an Outbreak Strain Causing *Mycoplasma bovis* Mastitis and Its Asymptomatic Carriage in the Herd: A Case Study from Idaho, USA. *Prev. Vet. Med.* **2010**, *93*, 66–70. [CrossRef]
30. Hurri, E.; Alvåsen, K.; Widgren, S.; Ohlson, A.; Aspán, A.; Pedersen, K.; Tråvén, M. A Longitudinal Study of the Dynamics of *Mycoplasma bovis* Antibody Status in Primiparous Cows and Bulk Tank Milk in Swedish Dairy Herds. *J. Dairy Sci.* **2024**. [CrossRef]
31. Burnens, A.; Bonnemain, P.; Bruderer, U.; Schalch, L.; Audigé, L.; Grand, D.; Poumarat, F.; Nicolet, J. The Seroprevalence of *Mycoplasma bovis* in Lactating Cows in Switzerland, Particularly in the Republic and Canton of Jura. *Schweiz. Arch. Tierheilkd.* **1999**, *141*, 455–460.
32. Teagasc National Farm Survey. Available online: https://www.teagasc.ie/media/website/publications/2022/Teagasc-National-Farm-Survey-2021.pdf (accessed on 7 September 2024).

33. Gonzalez, R.N.; Sears, P.M.; Merrill, R.A.; Hayes, G.L. Mastitis Due to Mycoplasma in the State of New York during the Period 1972–1990. *Cornell Vet.* **1992**, *82*, 29–40. [PubMed]
34. Valle, P.; Martin, S.; Tremblay, R.; Bateman, K. Factors Associated with Being a Bovine-Virus Diarrhoea (BVD) Seropositive Dairy Herd in the Møre and Romsdal County of Norway. *Prev. Vet. Med.* **1999**, *40*, 165–177. [CrossRef] [PubMed]
35. Correa-Valencia, N.M.; Ramírez, N.F.; Arango-Sabogal, J.C.; Fecteau, G.; Fernández-Silva, J.A. Prevalence of *Mycobacterium avium* subsp. *paratuberculosis* Infection in Dairy Herds in Northern Antioquia (Colombia) and Associated Risk Factors Using Environmental Sampling. *Prev. Vet. Med.* **2019**, *170*, 104739. [CrossRef] [PubMed]
36. Gabinaitienė, A.; Šiugždaitė, J.; Žilinskas, H.; Šiugžda, R.; Petkevičius, S. *Mycoplasma bovis* and Bacterial Pathogens in the Bovine Respiratory Tract. *Veterinární Medicína* **2011**, *56*, 29–35. [CrossRef]
37. Nicholas, R.; Ayling, R.; Stipkovits, L. An Experimental Vaccine for Calf Pneumonia Caused by *Mycoplasma bovis*. *Vaccine* **2002**, *20*, 3569–3575. [CrossRef]
38. Byrne, W.; Markey, B.; McCormack, R.; Egan, J.; Ball, H.; Sachse, K. Persistence of *Mycoplasma bovis* Infection in the Mammary Glands of Lactating Cows Inoculated Experimentally. *Vet. Rec.* **2005**, *156*, 767–771. [CrossRef]
39. Gille, L.; Evrard, J.; Callens, J.; Supré, K.; Grégoire, F.; Boyen, F.; Haesebrouck, F.; Deprez, P.; Pardon, B. The Presence of *Mycoplasma bovis* in Colostrum. *Vet. Res.* **2020**, *51*, 54. [CrossRef]
40. Timsit, E.; Arcangioli, M.-A.; Bareille, N.; Seegers, H.; Assié, S. Transmission Dynamics of *Mycoplasma bovis* in Newly Received Beef Bulls at Fattening Operations. *J. Vet. Diagn. Investig.* **2012**, *24*, 1172–1176. [CrossRef]
41. Godden, S.; Lombard, J.; Woolums, A. Colostrum Management for Dairy Calves. *Vet. Clin. N. Am. Food Anim. Pract.* **2019**, *35*, 535–556. [CrossRef]
42. Raboisson, D.; Trillat, P.; Cahuzac, C. Failure of Passive Immune Transfer in Calves: A Meta-Analysis on the Consequences and Assessment of the Economic Impact. *PLoS ONE* **2016**, *11*, e0150452. [CrossRef]
43. Vasseur, E.; Borderas, F.; Cue, R.I.; Lefebvre, D.; Pellerin, D.; Rushen, J.; Wade, K.M.; de Passillé, A.M. A Survey of Dairy Calf Management Practices in Canada That Affect Animal Welfare. *J. Dairy Sci.* **2010**, *93*, 1307–1316. [CrossRef]
44. Hughes, G.; Choudhury, R.; McRoberts, N. Summary Measures of Predictive Power Associated with Logistic Regression Models of Disease Risk. *Phytopathology* **2019**, *109*, 712–715. [CrossRef]

animals

MDPI

Article

First Molecular Detection and Genetic Characterization of *Tetratrichomonas buttreyi* and *Pentatrichomonas hominis* in Donkeys in Shanxi Province, China

Han-Dan Xiao [1], Shuo Zhang [1], Yi-Han Lv [1], Ze-Dong Zhang [1], Nan Su [1], Liang-Liang Li [2], Xing-Quan Zhu [1], Shi-Chen Xie [1,*] and Wen-Wei Gao [1,*]

1 Laboratory of Parasitic Diseases, College of Veterinary Medicine, Shanxi Agricultural University, Jinzhong 030801, China; 15166873600@163.com (H.-D.X.); shuoshuozhang0828@163.com (S.Z.); yh001028@126.com (Y.-H.L.); zzd18203541696@163.com (Z.-D.Z.); sunan1228@163.com (N.S.); xingquanzhu1@hotmail.com (X.-Q.Z.)

2 Taiyuan Zoo, Taiyuan 030009, China; liangliang20132017@163.com

* Correspondence: xieshichen221@163.com (S.-C.X.); sxndgaowenwei@163.com (W.-W.G.)

Simple Summary: Trichomonads are among the most prevalent intestinal parasites with a worldwide distribution which can infect many animals, resulting in economic losses and threatening public health. The donkey raising industry in Shanxi Province is relatively well-developed; however, it is not yet known whether donkeys in Shanxi Province were infected with *Tetratrichomonas buttreyi* and *Pentatrichomonas hominis*. Thus, 815 fecal samples were collected from donkeys in three representative geographical locations in Shanxi Province to determine the prevalence and associated risk factors of *T. buttreyi* and *P. hominis* in donkeys using molecular approaches. The overall prevalence of *T. buttreyi* and *P. hominis* in donkeys in Shanxi Province was 25.4% and 0.7%, respectively. Genetic analysis revealed that all *P. hominis* sequences obtained in this study were identified as genotype CC1, suggesting possible zoonotic potential. This is the first report of *T. buttreyi* and *P. hominis* prevalence in donkeys worldwide, which not only extends the geographical distribution of trichomonads but also expands the host spectrum. The findings also have implications for the prevention and control of trichomonad infections in donkeys in Shanxi Province.

Abstract: Two species of trichomonads, *Tetratrichomonas buttreyi* and *Pentatrichomonas hominis*, are common intestinal parasites that can impact animal health and productivity. Severe infection by these parasites can lead to diarrhea and wasting in affected animals. Notably, *P. hominis* is known to cause diarrhea and has the potential to be transmitted between animals and humans. Donkeys hold significant economic importance in China's agricultural sector. However, whether donkeys are infected with *T. buttreyi* and *P. hominis* remains unknown globally. To address this gap in knowledge, 815 fecal samples were collected from donkeys in three representative regions in Shanxi Province, North China. Then, the presence and genetic characteristics of *T. buttreyi* and *P. hominis* were examined using species-specific PCR primers amplifying the small subunit ribosomal RNA genes. The overall prevalence was detected to be 25.4% (207/815) for *T. buttreyi* and 0.7% (6/815) for *P. hominis* in donkeys in Shanxi Province. All obtained *P. hominis* sequences were identified as genotype CC1. Genetic analysis revealed that all *P. hominis* isolates from donkeys were clustered into the same branch with isolates detected in humans, suggesting possible zoonotic transmission. This study is the first to report the occurrence and prevalence of *T. buttreyi* and *P. hominis* in donkeys globally. These findings expand the host range of trichomonads and improve our understanding of their genetic diversity and zoonotic potential, providing essential baseline data for the prevention and control of these parasites in donkeys in the region.

Keywords: *Tetratrichomonas buttreyi*; *Pentatrichomonas hominis*; donkey; prevalence; zoonotic parasites; Shanxi Province

1. Introduction

Tetratrichomonas buttreyi and *Pentatrichomonas hominis* are two protozoan parasites of the Trichomonadidae family that inhabit the gastrointestinal tracts of humans and animals as parasites or commensals, posing significant public health challenges [1]. They exist in a trophozoite form, which is responsible for infection and replication within the intestines [2]. Both parasites have direct life cycles, with transmission occurring primarily through fecal–oral routes, and exhibit distinct characteristics and implications for human health [3,4].

In 1960, *T. buttreyi* was first identified in the ceca of pigs by Hibler et al. [5] and was considered a non-pathogenic commensal organism detected in pigs and cattle [2,6]. Recently, a report indicated that excessive infection by trichomonads can be pathogenic [7], and subsequently, symptoms such as diarrhea were observed in dairy cattle which were infected with *T. buttreyi* [4].

Existing evidence indicates that *P. hominis* is an opportunistic parasite causing diarrhea in humans, monkeys, dogs, pigs and cattle [8–12]. In addition, previous studies have shown that *P. hominis* may be recognized as a causative agent of diarrhea with potential for zoonotic transmission [13,14]. To date, most reports of *P. hominis* involve canids, with the prevalent genotypes detected in dogs being CC1, CC2 and CC3 [15]. It has also been sporadically reported in humans [16,17]. However, a previous study demonstrated that *P. hominis* infections may accelerate the development of colon cancer through changing gut microbiota [14]. With the deeper understanding of *P. hominis*, an increasing number of reports have indicated that *P. hominis* not only reproduces at the cecum or colon, but has also been detected in other organs, such as the anocelia [18,19].

Typically, microscopic examination is the routine method to discriminate trichomonad species. However, it is difficult to distinguish trichomonads due to their similar morphology under the microscope (e.g., *Trichomonas foetus* and *T. buttreyi*). With the rapid development of molecular detection methods, polymerase chain reaction (PCR)-based approaches have become important tools for detecting and identifying the trichomonads with higher specificity and sensitivity, especially in asymptomatic individuals [20–22]. The small subunit ribosomal RNA (SSU rRNA) gene is the main genetic marker to identify the species and genotypes of trichomonads [12].

The accurate identification of various trichomonad species is important for the diagnosis, treatment and surveillance of trichomonad infections in humans and animals. Ronidazole is a potentially neurotoxic drug, used for the treatment of feline trichomoniasis caused by *Trichomonas foetus* infection [23]. Metronidazole is considered the drug of choice for the treatment of *P. hominis*; however, it is proven to be ineffective against *Trichomonas foetus* [3]. Therefore, the accurate identification of trichomonad species is necessary to establish the correct treatment plan.

China is among the top countries in donkey breeding in the world. Historically, donkeys have been valuable for trade and are now appreciated for their nutritional benefits [24]. Donkeys play a significant economic role in rural areas, providing tender meat, nutritious skin, and milk [25–27]. Due to the growing significance of trichomonads in veterinary medicine, an increasing number of studies have been conducted on the prevalence and pathogenicity of trichomonad infections in different vertebrates. However, no studies have been published on the epidemiology of *T. buttreyi* and *P. hominis* in donkeys globally. Thus, this study firstly investigated the occurrence, prevalence and genetic characterization of *T. buttreyi* and *P. hominis* in donkeys in Shanxi Province, expanding the host spectrum and providing the baseline data to control and prevent these parasites in the study areas.

2. Materials and Methods

2.1. Sampling Collection

From April to May 2023, 815 fresh fecal specimens were sampled from donkeys in three representative cities in Shanxi Province: 81 from Jinzhong city, 363 from Linfen city and 371 from Datong city. To minimize contamination, the uppermost part of each freshly excreted fecal sample was collected using a disposable glove and recorded with relevant details, including region, sex and age. The donkey feces were categorized into two age groups (donkeys aged three years and above, and those which were lower than 3 years) and two sex groups (male and female). All samples were then transported under cool conditions to the Laboratory of Parasitic Diseases, College of Veterinary Medicine, Shanxi Agricultural University, and they were stored at −20 °C until needed for PCR-based molecular analysis.

2.2. DNA Extraction and PCR Amplification

Following the instructions provided in the E.Z.N.A.® Stool DNA Kit (Omega Biotek, Inc., Norcross, GA, USA), genomic DNA was extracted from approximately 200 mg of each fecal sample and then stored at −20 °C until PCR amplification. A total of 25 μL PCR mixture was prepared, including 2 μL of dNTPs, 2.5 μL of 10× PCR Buffer (Mg^{2+} free), 25 mM of $MgCl_2$, 1.25 U of *Ex*-Taq (Takara, Dalian, China), 1 μL of each primer, 2 μL of genomic DNA and 14.75 μL of ddH_2O. The PCR primers and amplification procedures referred to previous studies [8,28] and are listed in Table 1. Each PCR assay included both negative controls (reagent-grade water) and positive controls (verified DNA of *T. buttreyi* or *P. hominis* by sequencing) to ensure the reliability of the results. The amplicons were analyzed on 1.5% agarose gels containing ethidium bromide and observed using UV transillumination, and the positive ones were sequenced by Sangon Biotech Co., Ltd. (Shanghai, China) bidirectionally.

Table 1. PCR primers and parameters used in this study.

Species	Gene	Primer ID	Primer Sequences (5′-3′)	Annealing Temperatures (°C)	Fragment Length (bp)
T. buttreyi	SSU rRNA	FF	GCGCCTGAGAGATAGCGACTA	59	
		RR	GGACCTGTTATTGCTACCCTCTTC		
		bF	GTTTTTTCTCAGGCAGCAATG	61	623
		bR	GCAACCTAGAAACCTAGGCG		
P. hominis	SSU rRNA	F1	ATGGCGAGTGGTGGAATA	60	
		R1	CCCAACTACGCTAAGGATT		
		F2	TGTAAACGATGCCGACAGAG	60	339
		R2	CAACACTGAAGCCAATGCGAGC		

2.3. Sequencing and Phylogenetic Analysis

In this study, we utilized Chromas V2.6 software to proofread and assemble the obtained sequences; then, the Basic Local Alignment Search Tool (BLAST) was subsequently used to identify species by alignment with relevant sequences of known species available in the GenBank database. A phylogenetic analysis was conducted with the Neighbor-joining (NJ) method in MEGA 7.0 software, applying the Kimura-2-parameter model. To evaluate the robustness of the reconstructed phylogenetic trees, we performed a bootstrap analysis with 1000 replicates.

2.4. Statistical Analysis

The chi-square ($\chi 2$) test was used to evaluate the relevance between the prevalence of *T. buttreyi* or *P. hominis* across various regions, ages and sexes, employing SPSS 26.0 software (SPSS Inc., Chicago, IL, USA). Moreover, odds ratios (ORs) and 95% confidence intervals (95% CIs) were calculated to determine the strength of the correlation between prevalence and the examined variables. A *p*-value of less than 0.05 was considered to be statistically significant.

3. Results

3.1. Prevalence of T. buttreyi and P. hominis in Donkeys

In this study, 207 out of 815 fecal samples and 6 out of 815 fecal samples from donkeys were detected as *T. buttreyi*- and *P. hominis*-positive, respectively. The overall prevalence in Shanxi Province was 25.4% for *T. buttreyi* (95% CI: 22.4–28.4) and 0.7% for *P. hominis* (95% CI: 0.2–1.3), respectively (Table 2). Among the donkeys in the three cities examined, donkeys in Linfen city had the highest *T. buttreyi* prevalence of 31.7% (115/363), while donkeys in Datong city had the highest *P. hominis* prevalence of 1.1% (4/371). Statistical analysis showed that significant differences in the prevalence of *T. buttreyi* were observed in donkeys among region groups ($p < 0.001$) and sex groups ($p < 0.001$). In contrast to *T. buttreyi*, no statistically significant difference was found in donkeys between region groups and sex groups in the prevalence of *P. hominis* ($p > 0.05$). However, a statistically significant difference in *P. hominis* prevalence ($p < 0.05$) was found between donkeys aged ≥3 years (0.3%, 2/601) and donkeys aged <3 years (1.9%, 4/214). Additionally, among the 815 fecal samples, the co-infection of both *T. buttreyi* and *P. hominis* was detected in a female donkey in Datong city which was aged less than 3 years, with no clinical symptoms.

Table 2. Factors associated with prevalence of *T. buttreyi* and *P. hominis* in donkeys in Shanxi Province, North China.

Species	Factor	Category	No. Positive/No. Tested	Prevalence % (95% CI)	OR (95% CI)	*p*-Value
T. buttreyi	Region	Jinzhong	16/81	19.8 (11.1–28.4)	Ref.	<0.001
		Linfen	115/363	31.7 (26.9–36.5)	1.9 (1.0–3.4)	
		Datong	76/371	20.5 (16.4–24.6)	1.1 (0.6–1.9)	
	Age	≥3 years	160/601	26.6 (23.1–30.2)	1.3 (0.9–1.9)	0.179
		<3 years	47/214	22.0 (16.4–27.5)	Ref.	
	Sex	Male	15/120	12.5 (6.6–18.4)	Ref.	<0.001
		Female	192/695	27.6 (24.3–31.0)	2.7 (1.5–4.7)	
		Sub-total	207/815	25.4 (22.4–28.4)		
P. hominis	Region	Jinzhong	0/81	0		0.428
		Linfen	2/363	0.6 (0.0–1.3)	Ref.	
		Datong	4/371	1.1 (0.0–2.1)	2.0 (0.4–10.8)	
	Age	≥3 years	2/601	0.3 (0.0–0.8)	Ref.	0.024
		<3 years	4/214	1.9 (0.1–3.7)	5.7 (1.0–31.4)	
	Sex	Male	2/120	1.7 (0.0–4.0)	2.9 (0.5–16.2)	0.197
		Female	4/695	0.6 (0.0–1.1)	Ref.	
		Sub-total	6/815	0.7 (0.2–1.3)		

3.2. Sequence Analysis of T. buttreyi *and* P. hominis

T. buttreyi-positive samples were sequenced, and nine distinct sequence types showing 98.1–99.8% sequence similarity were obtained. Among the 207 *T. buttreyi* sequences obtained from donkeys in this study, 129, 64 and 8 sequences were identical to the reported *T. buttreyi* sequences in China with accession numbers PP256577 (pig), PP256576 (pig) and MK880285 (cattle), respectively. Six other sequences showed 98.4–99.8% identity to the reported *T. buttreyi* sequence (accession number: PP256576) isolated from pigs in Shanxi Province.

Regarding the obtained 6 sequences of *P. hominis* in this study, comparative analysis showed that 66.7% (4/6) of these sequences had 100% similarity to the reported *P. hominis* sequence isolated from a fox in China (accession number: OM763804), and another 2 sequences exhibited 99.7% homology with reference sequences isolated from dogs in China (KX136890 and KX136876), respectively. In addition, all *P. hominis* sequences obtained from donkeys in this study were identified as genotype CC1.

3.3. Phylogenetic Analysis of T. buttreyi *and* P. hominis

To better understand the genetic relationship of *T. buttreyi* and *P. hominis* detected in this study, a phylogenetic tree was reconstructed including other related trichomonad species. As shown in Figure 1, sequences of *T. buttreyi* and *P. hominis* from this study were clustered with reported animal-derived sequences. Notably, the three representative sequences of *P. hominis* from donkeys also clustered with a *P. hominis* sequence isolated from a human, indicating potential zoonotic transmission. The representative sequences from this study were deposited in the GenBank database with the following accession numbers: PQ113556 to PQ113564 for *T. buttreyi* and PQ114251 to PQ114253 for *P. hominis*.

Figure 1. Phylogenetic relationship of trichomonad species inferred from SSU rRNA gene sequences using Neighbor-joining analysis, based on Kimura two-parameter model, with 1000 bootstrap replications. *T. buttreyi* sequences obtained in this study are marked with black circle (•) and those of *P. hominis* (▲) are marked with black triangle. Bootstrap values are shown when >50%.

4. Discussion

T. buttreyi and *P. hominis* are parasitic protozoans that commonly inhabit the intestinal tracts of various vertebrates. Notably, *P. hominis* has been verified as a zoonotic parasite infecting a number of mammals such as humans, primates, cats, dogs and cattle, causing serious gastrointestinal symptoms [18,29,30]. The trophozoite stage of *P. hominis* can form a pseudocyst under adverse conditions and can survive outside the host for several days, thereby increasing the risk of infection to other hosts [10,30]. Up to now, no studies have reported the occurrence of *T. buttreyi* and *P. hominis* in donkeys globally. Thus, the present study first examined the occurrence and genetic characterization of *T. buttreyi* and *P. hominis* in donkeys.

In the present study, the prevalence of *T. buttreyi* in donkeys in Shanxi Province was 25.4% (207/815), which was higher than the average prevalence in cattle in China [31] and lower than that in pigs in other provinces of China [6] and some other countries, e.g., the

Philippines [32]. Interestingly, a recent study reported a significantly higher prevalence of *T. buttreyi* in pigs (49.7%, 180/362) in Shanxi Province [33]. These differences in *T. buttreyi* prevalence might be influenced by factors such as geographic location, animal species, age distribution, feeding and management practices, ecological conditions, sex composition and the immune status of the animals. Further studies sampling larger numbers of animals and diverse animal species are needed to better understand the factors influencing the prevalence of *T. buttreyi* in different animals.

Shanxi Province, characterized by a loess-covered mountainous plateau, experiences significant variations in precipitation due to its topography, with annual rainfall ranging from 358 to 621 mm [34]. The highest prevalence of *T. buttreyi* in donkeys in this study was observed in Linfen city, which is located in the southern part of Shanxi Province and has higher humidity compared to other cities. A previous report indicated that trichomonads can survive for several days in moist environments [1]. Thus, we speculate that the favorable temperature and humidity in Linfen city contribute to the higher prevalence of *T. buttreyi* in donkeys. Additionally, the prevalence of *T. buttreyi* in donkeys in this study showed an age-dependent increase, which is not consistent with a previous report in pigs in China [8]. Statistical analysis also showed significant differences in *T. buttreyi* prevalence among sex groups ($p < 0.001$), with female donkeys showing a higher prevalence. Also, sex has been identified as a risk factor for trichomonad infection in non-human primates in China [35].

Based on SSU rRNA gene sequences of *P. hominis*, the prevalence of *P. hominis* in donkeys in Shanxi Province was 0.7% (6/815, 95% CI: 0.2–1.3). Notably, *P. hominis* was found in all regions except Jinzhong city. With regard to the age groups, a statistically significant difference in the prevalence of *P. hominis* was observed in the examined donkeys, and donkeys aged <3 years had a 5.7 times higher risk of infection compared with those aged ≥3 years. Previous studies also suggest that age is a critical factor in *P. hominis* transmission among animals and humans, but more epidemiological investigations are required to reveal the risk factors affecting the prevalence of *P. hominis* infection in different hosts, and to elucidate the pathogenic potential of *P. hominis* in young donkeys [31,36]. Generally, younger animals are more susceptible to parasites due to their less developed immune systems.

The gut microbiota, a complex ecosystem within the host, is essential for maintaining immune and metabolic homeostasis [37,38]. Studies have shown that infections with many gastrointestinal parasites often disrupt this balance, impacting host health [36]. *P. hominis* infection in female foxes, for instance, has been linked to gut microbiota imbalances, diarrhea and wasting symptoms [15]. Moreover, *P. hominis* can exacerbate colon cancer by altering patients' gut microbiota [14,39].

Close connections between hosts and through fecal–oral routes via the ingestion of trophozoites are considered routes of *P. hominis* transmission [36]. In recent years, *P. hominis* has been identified in the feces of felines and canids, and in economic animals such as cattle [30], pigs [8] and goats [40], suggesting that these animals can act as reservoirs for further transmission [41]. Overall, six sequences obtained in this study were identified as genotype CC1, which was frequently detected in canids, e.g., dogs, foxes and raccoon dogs [15,30]. Notably, the genotype CC1 was also reported in Siberian tigers (*Panthera tigris altaica*) [28], dogs [41], monkeys [36], goats [40], foxes [15] and humans [36] in China, indicating that this genotype is not host-specific and suggesting potential zoonotic transmission of *P. hominis* between different hosts. Dogs present on donkey farms may contribute to the *P. hominis* infection of donkeys, though the transmission between donkeys and dogs remains unclear. In addition, *P. hominis* has been detected in wild animals like the boa (*Boa constrictor imperator*) and the Philippine scops owl (*Otus megalotis*), suggesting its wide host spectrum and potential health risks to both humans and animals [32].

Phylogenetic analysis indicated that the six sequences of *P. hominis* obtained from donkeys in this study were clustered into one branch containing known *P. hominis* sequences identified in humans, suggesting potential zoonotic transmission. The present study used

SSU rRNA sequences as genetic markers for the identification of *T. buttreyi* and *P. hominis*. However, SSU rRNA sequences have limitations as genetic markers for the differentiation of closely related species and/or cryptic species [42,43]. Thus, more appropriate genetic markers, such as the internal transcribed spacers (ITS-1 and ITS-2) and mitochondrial cytochrome oxidase subunit I (*cox*1), should be used for the precise identification and accurate differentiation of closely related species and/or cryptic species [42,43]. Notably, no diarrhea symptoms were observed in the positive donkeys, and all had normal stool consistency. Therefore, further research is needed to confirm the pathogenicity of *P. hominis* infection in donkeys. This study not only addresses the knowledge gap of *T. buttreyi* and *P. hominis* infection in donkeys worldwide, but also provides useful information for implementing measures to control *T. buttreyi* and *P. hominis* infections in donkeys in the studied areas.

5. Conclusions

This study revealed that the prevalence of *T. buttreyi* and *P. hominis* in donkeys in Shanxi Province was 25.4% and 0.7%, respectively. Genetic analysis identified the CC1 genotype of *P. hominis* in these donkeys, suggesting that donkeys might serve as a potential host for *P. hominis* transmission. To our knowledge, this is the first report of the occurrence and prevalence of *T. buttreyi* and *P. hominis* in donkeys globally, which not only extends the host range of *T. buttreyi* and *P. hominis*, but also highlights the public health significance of *P. hominis*.

Author Contributions: Conceptualization, W.-W.G., S.-C.X. and X.-Q.Z.; methodology, S.-C.X., S.Z. and W.-W.G.; software, S.-C.X.; validation, Z.-D.Z.; formal analysis, H.-D.X.; investigation, H.-D.X., S.Z., Y.-H.L. and N.S.; resources, H.-D.X., S.Z., N.S. and L.-L.L.; data curation, H.-D.X., S.-C.X. and W.-W.G.; writing—original draft preparation, H.-D.X.; writing—review and editing, S.-C.X., W.-W.G. and X.-Q.Z.; visualization, H.-D.X.; supervision, S.-C.X., W.-W.G. and X.-Q.Z.; project administration, W.-W.G. and X.-Q.Z.; funding acquisition, X.-Q.Z. All authors have read and agreed to the published version of the manuscript.

Funding: This research was funded by Shanxi Province Excellent Doctoral Work Award-Scientific Research Project (Grant No. SXBYKY2021019), the Research Fund of Shanxi Province for Introduced High-level Leading Talents (Grant No. RFSXIHLT202101), the Fund for Shanxi "1331 Project" (Grant No. 20211331-13), and the Special Research Fund of Shanxi Agricultural University for High-level Talents (Grant No. 2021XG001). The funders had no role in the design of the study; in the collection, analyses, or interpretation of data; in the writing of the manuscript; or in the decision to publish the results.

Institutional Review Board Statement: Not applicable.

Informed Consent Statement: Not applicable.

Data Availability Statement: The data sets supporting the results of this article have been submitted to GenBank, and the accession number is shown in the article.

Conflicts of Interest: The authors declare no conflicts of interest.

References

1. Gookin, J.L.; Hanrahan, K.; Levy, M.G. The conundrum of feline trichomonosis. *J. Feline Med. Surg.* **2017**, *19*, 261–274. [CrossRef] [PubMed]
2. Mostegl, M.M.; Richter, B.; Nedorost, N.; Maderner, A.; Dinhopl, N.; Kulda, J.; Liebhart, D.; Hess, M.; Weissenböck, H. Design and validation of an oligonucleotide probe for the detection of protozoa from the order trichomonadida using chromogenic in situ hybridization. *Vet. Parasitol.* **2010**, *171*, 1–6. [CrossRef] [PubMed]
3. Bastos, B.F.; Brener, B.; de Figueiredo, M.A.; Leles, D.; Mendes-de-Almeida, F. *Pentatrichomonas hominis* infection in two domestic cats with chronic diarrhea. *JFMS Open. Rep.* **2018**, *4*, 2055116918774959. [PubMed]
4. Castella, J.; Muñoz, E.; Ferrer, D.; Gutiérrez, J.F. Isolation of the trichomonad *Tetratrichomonas buttreyi* (Hibler et al., 1960) Honigberg, 1963 in bovine diarrhoeic faeces. *Vet. Parasitol.* **1997**, *70*, 41–45. [CrossRef] [PubMed]
5. Rivera, W.L.; Lupisan, A.J.; Baking, J.M. Ultrastructural study of a tetratrichomonad isolated from pig fecal samples. *Parasitol. Res.* **2008**, *103*, 1311–1316. [CrossRef]

6. Li, W.C.; Wang, K.; Li, Y.; Zhao, L.P.; Xiao, Y.; Gu, Y.F. Survey and molecular characterization of trichomonads in pigs in Anhui Province, East China, 2014. *Iran. J. Parasitol.* **2018**, *13*, 602–610.
7. Mostegl, M.M.; Richter, B.; Nedorost, N.; Maderner, A.; Dinhopl, N.; Weissenböck, H. Investigations on the prevalence and potential pathogenicity of intestinal trichomonads in pigs using in situ hybridization. *Vet. Parasitol.* **2011**, *178*, 58–63. [CrossRef]
8. Li, W.; Li, W.; Gong, P.; Zhang, C.; Yang, J.; Zhang, X.; Li, J. The prevalence of intestinal trichomonads in Chinese pigs. *Vet. Parasitol.* **2015**, *211*, 12–15. [CrossRef]
9. Cobo, E.R.; Corbeil, L.B.; Agnew, D.W.; VanHoosear, K.; Friend, A.; Olesen, D.R.; BonDurant, R.H. *Tetratrichomonas* spp. and *Pentatrichomonas hominis* are not persistently detectable after intravaginal inoculation of estrous heifers. *Vet. Parasitol.* **2007**, *150*, 18–26. [CrossRef]
10. Meloni, D.; Mantini, C.; Goustille, J.; Desoubeaux, G.; Maakaroun-Vermesse, Z.; Chandenier, J.; Gantois, N.; Duboucher, C.; Fiori, P.L.; Dei-Cas, E.; et al. Molecular identification of *Pentatrichomonas hominis* in two patients with gastrointestinal symptoms. *J. Clin. Pathol.* **2011**, *64*, 933–935. [CrossRef]
11. Fukushima, T.; Mochizuki, K.; Yamazaki, H.; Watanabe, Y.; Yamada, S.; Aoyama, T.; Sakurai, Y.; Mori, H.; Nakazawa, M. *Pentatrichomonas hominis* from beagle dogs—Detection method, characteristics and route of infection. *Exp. Anim.* **1990**, *39*, 187–192. [CrossRef]
12. Gookin, J.L.; Stauffer, S.H.; Levy, M.G. Identification of *Pentatrichomonas hominis* in feline fecal samples by polymerase chain reaction assay. *Vet. Parasitol.* **2007**, *145*, 11–15. [CrossRef] [PubMed]
13. Abdo, S.M.; Ghallab, M.M.I.; Elhawary, N.M.; Elhadad, H. *Pentatrichomonas hominis* and other intestinal parasites in school-aged children: Coproscopic survey. *J. Parasit. Dis.* **2022**, *46*, 896–900. [CrossRef] [PubMed]
14. Zhang, H.; Yu, Y.; Li, J.; Gong, P.; Wang, X.; Li, X.; Cheng, Y.; Yu, X.; Zhang, N.; Zhang, X. Changes of gut microbiota in colorectal cancer patients with *Pentatrichomonas hominis* infection. *Front. Cell. Infect. Microbiol.* **2022**, *12*, 961974. [CrossRef] [PubMed]
15. Song, P.; Guo, Y.; Zuo, S.; Li, L.; Liu, F.; Zhang, T.; Dai, H.; Dong, H. Prevalence of *Pentatrichomonas hominis* in foxes and raccoon dogs and changes in the gut microbiota of infected female foxes in the Hebei and Henan Provinces in China. *Parasitol. Res.* **2023**, *123*, 74. [CrossRef] [PubMed]
16. Zhang, N.; Zhang, H.; Yu, Y.; Gong, P.; Li, J.; Li, Z.; Li, T.; Cong, Z.; Tian, C.; Liu, X.; et al. High prevalence of *Pentatrichomonas hominis* infection in gastrointestinal cancer patients. *Parasit. Vectors* **2019**, *12*, 423. [CrossRef]
17. Phuanukoonnon, S.; Michael, A.; Kirarock, W.S.; Pomat, W.S.; van den Biggelaar, A.H. Intestinal parasitic infections and anaemia among pregnant women in the highlands of Papua New Guinea. *Papua New Guin. Med. J.* **2013**, *56*, 119–125.
18. Maritz, J.M.; Land, K.M.; Carlton, J.M.; Hirt, R.P. What is the importance of zoonotic trichomonads for human health? *Trends Parasitol.* **2014**, *30*, 333–341. [CrossRef]
19. Dong, N.; Jin, X.; Huang, J.; Chen, K.; Li, Y.; Chen, C.; Hu, D.; Xie, Y. Tetratrichomonas in pyopneumothorax. *Am. J. Emerg. Med.* **2019**, *37*, 1215.e1–1215.e4. [CrossRef]
20. Cobo, E.R.; Campero, C.M.; Mariante, R.M.; Benchimol, M. Ultrastructural study of a tetratrichomonad species isolated from prepucial smegma of virgin bulls. *Vet. Parasitol.* **2003**, *117*, 195–211. [CrossRef]
21. Baltrušis, P.; Höglund, J. Digital PCR: Modern solution to parasite diagnostics and population trait genetics. *Parasit. Vectors* **2023**, *16*, 143. [CrossRef]
22. Maurer, J.J. Rapid detection and limitations of molecular techniques. *Annu. Rev. Food. Sci. Technol.* **2011**, *2*, 259–279. [CrossRef]
23. Gookin, J.L.; Copple, C.N.; Papich, M.G.; Poore, M.F.; Stauffer, S.H.; Birkenheuer, A.J.; Twedt, D.C.; Levy, M.G. Efficacy of ronidazole for treatment of feline *Tritrichomonas foetus* infection. *J. Vet. Intern. Med.* **2006**, *20*, 536–543. [CrossRef]
24. Davis, E. Donkey and mule welfare. *Vet. Clin. N. Am. Equine. Pract.* **2019**, *35*, 481–491. [CrossRef] [PubMed]
25. Seyiti, S.; Kelimu, A. Donkey industry in China: Current aspects, suggestions and future challenges. *J. Equine Vet. Sci.* **2021**, *102*, 103642. [CrossRef] [PubMed]
26. Wang, X.; Wang, T.; Liang, H.; Wang, L.; Akhtar, F.; Shi, X.; Ren, W.; Huang, B.; Kou, X.; Chen, Y.; et al. A novel SNP in NKX1-2 gene is associated with carcass traits in Dezhou donkey. *BMC Genom. Data* **2023**, *24*, 41. [CrossRef] [PubMed]
27. Yu, M.; Zhang, X.; Yan, J.; Guo, J.; Zhang, F.; Zhu, K.; Liu, S.; Sun, Y.; Shen, W.; Wang, J. Transcriptional specificity analysis of testis and epididymis tissues in Donkey. *Genes* **2022**, *13*, 2339. [CrossRef] [PubMed]
28. Zhang, H.; Zhang, N.; Gong, P.; Cheng, S.; Wang, X.; Li, X.; Hou, Z.; Liu, C.; Bi, T.; Wang, B.; et al. Prevalence and molecular characterization of *Pentatrichomonas hominis* in Siberian tigers (*Panthera tigris altaica*) in northeast China. *Integr. Zool.* **2022**, *17*, 543–549. [CrossRef]
29. Bailey, N.P.; Velo-Rego, E.; Hirt, R.P. Sporadic isolation of tetratrichomonas species from the cattle urogenital tract. *Parasitology* **2021**, *148*, 1339–1344. [CrossRef]
30. Mahittikorn, A.; Udonsom, R.; Koompapong, K.; Chiabchalard, R.; Sutthikornchai, C.; Sreepian, P.M.; Mori, H.; Popruk, S. Molecular identification of *Pentatrichomonas hominis* in animals in central and western Thailand. *BMC Vet. Res.* **2021**, *17*, 203. [CrossRef]
31. Li, W.C.; Huang, J.M.; Fang, Z.; Ren, Q.; Tang, L.; Kan, Z.Z.; Liu, X.C.; Gu, Y.F. Prevalence of *Tetratrichomonas buttreyi* and *Pentatrichomonas hominis* in yellow cattle, dairy cattle, and water buffalo in China. *Parasitol. Res.* **2020**, *119*, 637–647. [CrossRef] [PubMed]
32. Dimasuay, K.G.; Rivera, W.L. Molecular characterization of trichomonads isolated from animal hosts in the Philippines. *Vet. Parasitol.* **2013**, *196*, 289–295. [CrossRef]

33. Wang, Z.R.; Fan, Q.X.; Wang, J.L.; Zhang, S.; Wang, Y.X.; Zhang, Z.D.; Gao, W.W.; Zhu, X.Q.; Liu, Q. Molecular identi-fication and survey of Trichomonad species in pigs in Shanxi Province, North China. *Vet. Sci.* **2024**, *11*, 203. [CrossRef] [PubMed]
34. Gao, W.W.; Zhang, S.; Zhang, T.H.; Xiao, H.D.; Su, N.; Tao, M.F.; Wu, Z.X.; Zhang, Z.D.; Zhu, X.Q.; Xie, S.C. Prevalence and multilocus genotyping of *Giardia duodenalis* in donkeys in Shanxi Province, North China. *Animals* **2023**, *13*, 3771. [CrossRef] [PubMed]
35. Ma, P.P.; Zou, Y.; Mu, W.J.; Zhang, Y.Y.; Li, Y.Q.; Liu, Z.L.; Zhang, L.; Chen, L.X.; Liu, G.H.; Wang, S. Prevalence of in-testinal trichomonads in captive non-human primates in China. *Parasite* **2024**, *31*, 19. [CrossRef] [PubMed]
36. Li, W.C.; Ying, M.; Gong, P.T.; Li, J.H.; Yang, J.; Li, H.; Zhang, X.C. *Pentatrichomonas hominis*: Prevalence and molecular characterization in humans, dogs, and monkeys in Northern China. *Parasitol. Res.* **2016**, *115*, 569–574. [CrossRef] [PubMed]
37. Peachey, L.E.; Castro, C.; Molena, R.A.; Jenkins, T.P.; Griffin, J.L.; Cantacessi, C. Dysbiosis associated with acute helminth infections in herbivorous youngstock - observations and implications. *Sci. Rep.* **2019**, *9*, 11121. [CrossRef]
38. Thursby, E.; Juge, N. Introduction to the human gut microbiota. *Biochem. J.* **2017**, *474*, 1823–1836. [CrossRef]
39. Dheilly, N.M.; Ewald, P.W.; Brindley, P.J.; Fichorova, R.N.; Thomas, F. Parasite-microbe-host interactions and cancer risk. *PLoS Pathog.* **2019**, *15*, e1007912. [CrossRef]
40. Li, W.C.; Wang, K.; Gu, Y. Occurrence of *Blastocystis* sp. and *Pentatrichomonas hominis* in sheep and goats in China. *Parasit. Vectors* **2018**, *11*, 93. [CrossRef]
41. Li, W.C.; Gong, P.T.; Ying, M.; Li, J.H.; Yang, J.; Li, H.; Yang, Z.T.; Zhang, G.C.; Zhang, X.C. *Pentatrichomonas hominis*: First isolation from the feces of a dog with diarrhea in China. *Parasitol. Res.* **2014**, *113*, 1795–1801. [CrossRef] [PubMed]
42. Ghafar, A.; Abbas, G.; Beasley, A.; Bauquier, J.; Wilkes, E.J.A.; Jacobson, C.; McConnell, E.; El-Hage, C.; Carrigan, P.; Cudmore, L.; et al. Molecular diagnostics for gastrointestinal helminths in equids: Past, present and future. *Vet. Parasitol.* **2023**, *313*, 109851. [CrossRef] [PubMed]
43. Akhoundi, M.; Downing, T.; Votýpka, J.; Kuhls, K.; Lukeš, J.; Cannet, A.; Ravel, C.; Marty, P.; Delaunay, P.; Kasbari, M.; et al. Leishmania infections: Molecular targets and diagnosis. *Mol. Aspects Med.* **2017**, *57*, 1–29. [CrossRef] [PubMed]

animals MDPI

Communication

Bovine Parainfluenza Virus 3 and Bovine Respiratory Syncytial Virus: Dominant Viral Players in Bovine Respiratory Disease Complex among Serbian Cattle

Vesna Milićević [1,*], Sofija Šolaja [1], Dimitrije Glišić [1], Milan Ninković [1], Bojan Milovanović [1], Milan Đorđević [2], Snežana Ristevski [3], Filip Spasojević [4] and Miroljub Dačić [5]

1 Institute of Veterinary Medicine of Serbia, Janisa Janulisa 14, 11000 Belgrade, Serbia; sofija.solaja@nivs.rs (S.Š.); dimitrije.glisic@nivs.rs (D.G.); milan.ninkovic@nivs.rs (M.N.); bojan.milovanovic@nivs.rs (B.M.)
2 Veterinary Clinic "Mladenovac", Kralja Petra I 347, 11400 Mladenovac, Serbia; djordjevicvet@gmail.com
3 Delta Vet Med doo, Vladimira Popovića 8A, 11000 Belgrade, Serbia; snezana.ristevski@deltaagrar.rs
4 Al Dahra Srbija doo, Gorskih Jasenova 4, 11000 Belgrade, Serbia; filip.spasojevic@aldahra.com
5 Veterinary Specialized Institute "Jagodina", Boška Jovića 6, 35000 Jagodina, Serbia; vsijdacic@gmail.com
* Correspondence: vesna.milicevic@nivs.rs

Simple Summary: This study thoroughly investigated viral pathogens associated with bovine respiratory disease complex (BRDC) in Serbian cattle using serum and nasal swab samples. Conducted in 2024 across 65 randomly selected dairy farms in Serbia, excluding ones with vaccinated cattle, this study categorized the farms by their size: small, medium, and large. Serum samples from adult cattle were tested for antibodies against BVDV, BHV-1, BRSV, and BPIV3, while nasal swabs from respiratory-symptomatic animals were PCR-tested for viral genome detection. The results showed seropositivity for all four viruses on all of the farms, with BPIV3 being universally positive. Medium-sized and large farms exhibited higher levels of seropositivity for BRSV and BHV-1 compared to small farms ($p < 0.05$). Our true seroprevalence estimates were 84.29% for BRSV, 54.08% for BVDV, 90.61% for BHV-1, and 84.59% for BPIV3 at the animal level. A PCR analysis of the nasal swabs detected BRSV (20%), BHV-1 (1.7%), BVDV (8%), and BPIV3 (10.9%), with no Influenza D virus found. This study provides crucial insights into viral pathogen prevalence and circulation in Serbian cattle with BRDC, emphasizing the importance of surveillance and control measures to manage respiratory diseases in cattle populations.

Abstract: Bovine respiratory disease complex, a complex respiratory ailment in cattle, results from a combination of viral and bacterial factors, compounded by environmental stressors such as overcrowding, transportation, and adverse weather conditions. Its impact extends beyond mere health concerns, posing significant economic threats to the cattle industry. This study presents an extensive investigation into viral pathogens associated with BRDC in Serbian cattle, utilizing serum samples and nasal swabs. A cross-sectional study was conducted in 2024 across 65 randomly selected dairy farms in Serbia, excluding farms with vaccinated cattle. The farms were categorized by their livestock count: small ($\leq$50 animals), medium (51–200 animals), and large (>200 animals). Serum samples from adult cattle older than 24 months were tested for antibodies against BVDV, BHV-1, BRSV, and BPIV3. Nasal swab samples from the animals with respiratory signs were tested using PCR for viral genome detection. The results showed seropositivity for all four viruses across all of the farms, with BPIV3 exhibiting universal seropositivity. Medium-sized and large farms demonstrated higher levels of seropositivity for BRSV and BHV-1 compared to small farms ($p < 0.05$). Our true seroprevalence estimates at the animal level were 84.29% for BRSV, 54.08% for BVDV, 90.61% for BHV-1, and 84.59% for BPIV3. A PCR analysis of the nasal swabs revealed positive detections for BRSV (20%), BHV-1 (1.7%), BVDV (8%), and BPIV3 (10.9%). Influenza D virus was not found in any of the samples. This study provides critical insights into the prevalence and circulation of viral pathogens associated with BRDC in Serbian cattle, emphasizing the importance of surveillance and control measures to mitigate the impact of respiratory diseases in cattle populations.

Keywords: bovine respiratory disease complex; bovine viral diarrhoea virus; bovine herpesvirus-1; bovine respiratory syncytial virus; bovine parainfluenza virus 3; Influenza D virus; prevalence; Serbia

1. Introduction

Bovine respiratory disease complex (BRDC) refers to a multifactorial respiratory condition that affects cattle. It is caused by a combination of viral and bacterial pathogens, as well as environmental stressors such as overcrowding, transportation, and adverse weather conditions [1]. BRDC is a significant concern in the cattle industry as it can lead to substantial economic losses due to decreased productivity, high treatment costs [2], and a mortality rate that can reach 70% [3]. For example, even with yearly preventive vaccination, BRDC still imposes an estimated annual cost of approximately GBP 80 million on the UK economy [4]. Bovine respiratory syncytial virus (BRSV), bovine parainfluenza virus 3 (BPIV3), bovine viral diarrhoea virus (BVDV), bovine herpesvirus 1 (BHV-1), bovine adenovirus (BadV), and bovine coronavirus (BoCV) are recognized as the primary causes of respiratory illness in cattle [5]. Influenza D virus (IDV) was first identified in 2011 [6] and has since been confirmed in North America, Europe, East Asia, and Australia [7]. Although it typically causes mild symptoms, recent metagenomic analyses have shown a positive association between IDV and BRDC [8]. Viral infections further exacerbate the conditions conducive to bacterial infections. The damage to the upper respiratory tract and impaired mucociliary clearance enhance bacterial adhesion to virus-infected cells, facilitating their growth and colony formation. This damage progresses to the tracheal mucosa epithelium, enabling bacteria to penetrate deeper into the respiratory tract. Viruses also hinder the function of macrophages and neutrophil leukocytes, which are crucial for host immune responses and phagocytosis [9]. BRDC affects animals with a range of symptoms lasting up to five days. They include fever, lethargy, anorexia, coughing, nasal and ocular discharges, and, in severe cases, strenuous breathing, while bacterial pathogens can trigger an acute phase response with systemic symptoms, like fever, loss of appetite, and respiratory issues. Neonatal calf diarrhoea, which can occur with or without fever, is also linked to BRD and may emerge after significant damage to the intestinal submucosa [10]. Though biosecurity and antibiotics are also pillars, the control of BRDC relies on vaccines [11]. However, despite advancements, current BRD vaccines show limited efficacy as indicated by the development of clinical disease even in vaccinated animals due to factors like improper administration and storage, as well as challenges in vaccinating young calves [4]. BRDC remains a persistent and widespread challenge globally, with significant prevalence observed across Europe. In Serbia, governmental initiatives targeting BRDC are notably absent, leaving vaccination against the disease optional. Moreover, the dearth of comprehensive data concerning the prevalence and ramifications of BRDC within the country further exacerbates the challenge of effectively managing this complex. Therefore, this study sought to address this critical gap by estimating the seroprevalence of the viral infections deemed most significant within the BRDC complex.

2. Materials and Methods

This study was carried out using serum samples from the state's annual leukosis survey. Ethical approval or consent to participate was therefore not required.

2.1. Study Design and Sampling

Serum samples were obtained from annual enzootic bovine leukosis surveys, which include testing of all adult cattle older than 24 months. This cross-sectional study was conducted during January and February 2024, encompassing 65 randomly selected dairy farms across Serbia where vaccination was not practised. The locations of the sampled farms are presented in Figure 1.

Figure 1. The purple dots represent the location of sampled cattle farms.

These farms were classified based on their livestock count: small farms with up to 50 animals (n = 50), medium-sized farms ranging from 51 to 200 animals (n = 10), and large farms housing over 200 animals (n = 5). The number of farms and individuals surveyed was determined based on a 2-stage sampling design. The herd-level sample size was calculated with an anticipated prevalence of 10%, a confidence level of 95%, and a precision of 0.05%, alongside data on the total number of cattle farms in Serbia obtained from the Statistical Office of the Republic of Serbia. The number of animals to be tested was determined to ensure 95% probability of detecting at least one positive animal if the herd is infected, considering the farm's animal count, an assumed prevalence of 20%, a confidence level of 95%, and a precision of 0.05%. All cattle were subject to testing in small farms with up to ten animals. The nasal swabs were collected from animals exhibiting clinical signs of respiratory illness. Symptoms varied from mild to moderate, with no severe cases observed. In total, 175 nasal swabs were gathered and tested, along with 1000 serum samples comprising 400 from small farms, 300 from medium-sized ones, and 300 from large farms.

2.2. Antibody Detection Tests

Serum samples were analysed to identify antibodies against BVDV, BHV-1, BRSV, and BPIV3. Commercial ELISA kits were employed for assessing antibodies against BVDV (PrioCHECK™ Bovine BVDV Ab Plate Kit by Prionics, Schlieren, Switzerland), BHV-1 (ID Screen® IBR gB Competition by IDvet, Grabels, France), and BRSV (INgezim BRSV Compac by Ingenasa, Madrid, Spain). Sample preparation and testing procedures were carried out in accordance with the guidelines provided by the respective test manufacturers. The interpretation of test results was conducted individually, adhering to the specified cut-off values provided by the test producers. Specific antibodies against BPIV3 were assessed by hemagglutination inhibition (HI) using 0.5% chicken red blood cells and 4 hemagglutination units of SF4 strain of BPIV3 genotype A (ATCC-VR 281, American Bioresearch, Gaithersburg, MD, USA) propagated in MDBK cell line (ATCC CCL-22). The serum samples were heat-inactivated prior testing at 37 °C for 30 min. The antibody titre was determined as the reciprocal of the highest serum dilution that completely inhibited hemagglutination. The initial serum dilution was set at 1:4, with the seropositivity cutoff level established at 4.

2.3. PCR

Nasal swab samples were subjected to PCR testing to detect the presence of BVDV, BHV-1, BRSV, IDV, and BPIV3 genomes. Viral nucleic acids were extracted from the swabs using the IndiSpin Pathogen kit (Indical, Laipzig, Germany) according to the manufacturer's instructions, following immersion in 1ml of phosphate-buffered saline (PBS). Published primers for BVDV [12], BRSV [13], BHV-1 [14], IDV [15], and BPIV3 [16] were utilized for specific genome amplification (Supplementary material, Table S1). Real-time PCR was conducted using Luna® Universal qPCR Master Mix (NEB, Ipswich, MA, USA) for BHV-1, while real-time RT-PCR was performed using Luna Universal Probe One-Step RT-qPCR Kit (NEB, USA) for BVDV, BRSV, and PI3. Samples with Ct values below 40 were considered positive. Gel-based RT-PCR was applied for IDV detection using OneStep RT-PCR kit (Qiagen, Hilden, Germany).

2.4. Statistical Analysis

Descriptive statistical methods were used for the analysis of results. The true seroprevalence was calculated using https://epitools.ausvet.com.au/trueprevalence (accessed on 13 March 2024), using imperfect test and 95% confidence level.

3. Results

None of the 65 farms tested seronegative for all four viruses (Supplementary material, Table S2). All of the farms tested seropositive for BPIV3. Additionally, all medium-sized and large farms were seropositive for BRSV and BHV-1. Conversely, significantly lower levels of seropositivity were observed for BVDV ($p < 0.05$) and BHV-1 ($p < 0.05$) in small farms (Table 1) compared to large and medium-size farms, respectively.

Table 1. Seroprevalence of viral pathogens in Serbian cattle farms.

Farm Size	Number of Farms/ Number of Tested Animals	BRSV		BVDV		BHV-1		BPIV3	
		Seropositive Herds (number/%)	Seropositive Animals (number/%)	Seropositive Herds (number/%)	Seropositive Animals (number/%)	Seropositive Herds (number/%)	Seropositive Animals (number/%)	Seropositive Herds (number/%)	Seropositive Animals (number/%)
1–50 animals	50/400	40/80	331/82.75	8/16	197/49.25	10/20	325/81.25	50/100	379/94.75
51–200 animals	10/300	10/100	260/86.67	5/50	150/50	10/100	283/94.33	10/100	210/70
More than 200 animals	5/300	5/100	245/81.67	4/80	193/64.33	5/100	290/96.67	5/100	250/83.33
TOTAL	65/1000	55/84.61	836/83.6	17/26.15	540/54	25/38.46	898/89.8	65/100	839/83.9

The true seroprevalence at the animal level of BRSV was estimated at 84.29% (81.81–86.5%, with a CI of 95%). The true seroprevalence of BVDV is estimated to be 54.08%, (50.92–57.21%, with a CI of 95%). Regarding BHV-1, the true seroprevalence was 90.61% (88.54–92.37%, with a CI of 95%). Lastly, for BPIV3, the true seroprevalence was 84.59% (82.14–86.78%, with a CI of 95%).

Of the 175 nasal swabs collected, 20% ($n = 35$) tested positive for BRSV, 1.7% ($n = 3$) for BHV-1, 8% ($n = 14$) for BVDV, and 10.9% ($n = 19$) for BPIV3 (Supplementary material, Table S2, Figures S1–S4). The farm detection rates varied across different pathogens: BVDV ranged from 11.11% to 26.67%, BHV-1 was detected in 10% of cases, BRSV showed a detection rate between 50% and 100%, and BPIV3 ranged from 3.33% to 20% in terms of detection frequency. BRSV circulation was verified in seven of the farms, comprising two large and five small ones. BHV-1 was confirmed at a single large farm. BVDV circulation was identified in three farms: two large and one medium-sized. BPIV3 was detected in ten farms, evenly split between five large and five medium-sized ones. Influenza D virus was not detected in any of the tested farms. A statistically significant ($p < 0.05$) frequency of BVDV and BPIV3 detection was observed for the large and medium-sized farms compared to that of the small ones.

The simultaneous circulation of BRSV and BPIV3 was detected at two large farms, which were BVDV- and BHV-1-seropositive. BVDV and BPIV3 were concurrent at one large and one medium farm. The simultaneous circulation of BVDV, BHV-1, and BPIV3 was confirmed at one large farm.

4. Discussion

This study represents the first comprehensive investigation of viral pathogens associated with BRDC in Serbian cattle. Unlike European countries where eradication or control programs may be in place, Serbian farms grapple with BRDC without such support. Serbia's cattle structure is diverse, leading to varied outcomes, and the results are segmented based on farm size to reflect this diversity.

Although IDV has been reported to be circulating throughout Europe since 2012, the virus was not detected in any of the tested samples in this study. Previous reports have suggested that its seroprevalence could reach as high as 94.6% [17]. However, the positivity rate from nasal swab tests is usually less than 10% [17]. One limitation of this research was the lack of available serology tests that could accurately determine the prevalence of IDV infections in cattle in Serbia. Additionally, it is essential to consider the less-intensive cattle import market in Serbia compared to that in other EU countries, which may impact the spread of the virus and thus its absence in Serbia. It was shown that virus shedding occurs after the import of young cattle, which contributes to a broader diffusion in destination countries and facilitates viral spread through livestock trade [18]. Furthermore, although small farms in Serbia practice very low biosecurity measures, they tend to produce their own replacement cattle rather than purchasing them from larger farms, which can also limit the spread of IDV [19].

In contrast to IDV, BPIV3 emerged as the most prevalent respiratory viral pathogen in cattle in Serbia, irrespective of farm size. The seroprevalence of BPIV3 varies globally, reaching 100% at the herd level in places like Iran [20], and our study similarly found a herd seroprevalence of 100%. Furthermore, the BPIV3 genome was detected in 10.9% of nasal swabs in our study, aligning with previous research in Serbia, which reported a positivity rate of 6.7% using genome detection and virus isolation [21]. The average prevalence of BPIV3 in nasal swabs, as determined by PCR, was estimated at 7%, with age showing a significant influence but not farm type, which is consistent with our findings [22]. Considering the typical lack of control measures for BPIV3, even on large farms in Serbia, along with the virus's characteristics, such as efficient horizontal transmission, propensity for sub-clinical infections, and potential for reintroduction into herds [23], these results were anticipated.

Like BPIV3, BRSV also exhibited a trend toward 100% seroprevalence at the herd level in medium-sized and large farms. However, seroprevalence was notably lower in small farms, with 50% showing seronegativity. Among all of the viruses circulating in cattle, BRSV exhibits the highest pathogenicity, manifesting with clinical signs ranging from mild to moderate or even subclinical [24]. Despite the potential for cattle on farms to remain free from clinical symptoms due to the subclinical nature of the disease [25], this study highlighted that small farms could maintain seronegativity under extensive conditions. Nonetheless, efficient inter-herd transmission contributed to a high true seroprevalence of 84.29%. These findings echo other studies' findings, attributing the prevalence to year-round virus circulation [26] and repeated exposure leading to reinfections [27]. However, although the rate of positive nasal swabs from sick animals in our study was 20%, which was higher than those of other authors [28], it should be noted that viral RNA can be detected for up to 27 days [29], leading to an increased chance of detection.

While BPIV3 and BRSV are commonly reported in cattle, several countries or regions have achieved freedom from BoHV-1 through the implementation of eradication programs [30]. However, in Irish beef cattle, the herd-level seroprevalence to BHV-1 was as high as 90%, with a mean within-herd prevalence of 40% [31]. Consistent with findings in other studies [31,32], significant disparities in seroprevalence among small, medium-sized, and large farms were observed in this study. Reports from Estonia [32] indicated a substantial increase in herd prevalence with herd size, reaching 3.4% in the smallest category, consistent with previous studies from Serbia [33], while the mean within-herd prevalence was 37.8%, corresponding closely to our findings of 38.46%.

Moreover, seroprevalence tends to increase with age [33] due to latency and lifelong exposure to the virus. In this study, only animals aged 24 months and older were included in the seroprevalence estimation; thus, this association could not be confirmed. Contrary to our serological findings and owing to the nature of BHV-1, the virus itself was detected in only 1.7% of the nasal swabs from the sick animals. Similar observations were made in Slovenia, where the detection rate of BHV-1 was 0.75% [34]. However, considering age-related infection, the detection rate in young animals could be notably higher, as evidenced by findings from Poland, where the BHV-1 genome was detected in 36.5% of nasal swabs from young beef cattle [35].

While several European countries have successfully eradicated bovine viral diarrhoea virus (BVDV), leading to a reduced prevalence rate of 1.5% [36], this study reveals that BVDV seroprevalence remains high in Serbia and is particularly contingent on farm size. For herds with up to 50 animals, the seroprevalence was determined to be 16%, surpassing the prevalence observed in a study focused solely on backyard farms in the Belgrade area, where it was 3.8%, albeit with within-herd rates of up to 80% [19]. However, variations in prevalence are evident concerning region and farm management, but the within-herd prevalence usually approaches 100% [37]. Globally, the seropositivity rate stands at 42.77%, with dairy cattle demonstrating the highest prevalence at 48.68%. Notably, positive rates were more pronounced during summer (60.16%) and winter (63.44%), while cows exhibited a lower positivity rate compared to that of bulls, and calves showed a lower rate compared to that of adult cattle [36]. In comparison to antibodies, BVDV is considerably less detectable. In this study, only 8% of the nasal swabs from such animals contained BVDV, which is lower than the global average [36].

BPIV3 and BRSV were concurrently detected on two large farms, despite the presence of BVDV and BoHV-1, which are typically absent in order for BPIV3 and BRSV to become predominant viral pathogens, according to previous research [38]. A confirmation of the concurrent circulation of BVDV, BHV-1, and BPIV3 was obtained from a single large farm. The immunosuppression induced by BVDV infection facilitates other infections, resulting in a synergistic effect for numerous viral and bacterial pathogens responsible for respiratory illnesses.

Controlling the spread of BRDC in cattle populations is vital for their health and welfare. Key policies and practices include strict biosecurity measures to prevent infectious agent transmission, effective vaccination programs targeting common BRDC pathogens like BVDV, BHV-1, and BRSV, regular diagnostic testing to identify carriers, and implementing good management practices to reduce stress and minimize BRDC risk. These measures collectively aim to reduce disease prevalence and severity, ensuring healthier and more resilient cattle populations.

The study underscores the endemic nature and complex dynamics of viral pathogens associated with BRDC in Serbian cattle, influenced by factors such as farm size. Further research and targeted control measures are needed to mitigate the impact of these diseases on cattle health and productivity in Serbia.

5. Conclusions

This study emphasizes that viral pathogens linked to BRDC in Serbian cattle are not only prevalent but also exhibit intricate interactions, indicating an endemic nature.

These pathogens, including BVDV, BHV-1, BRSV, and BPIV3, play a significant role in the occurrence and severity of BRDC. Importantly, the dynamics of these pathogens are influenced by various factors, with farm size being a notable factor. This underscores the importance of considering farm size as a factor in understanding and addressing BRDC in Serbian cattle herds. Further research is warranted to delve deeper into these complex dynamics, including investigating how farm management practices and environmental factors may also contribute to disease prevalence and transmission. Moreover, this study highlights the need for targeted control measures to effectively mitigate the impact of these diseases on cattle health and productivity in Serbia. This includes implementing vaccination programs, improving biosecurity measures, and enhancing surveillance efforts to monitor disease prevalence and detect outbreaks early. By better understanding the endemic nature and dynamics of these viral pathogens and implementing appropriate control strategies, it is possible to reduce the burden of BRDC on Serbian cattle herds, ultimately improving animal welfare and economic outcomes for cattle producers.

Supplementary Materials: The following supporting information can be downloaded at: https://www.mdpi.com/article/10.3390/ani14101458/s1, Figure S1: qPCR_BHV-1; Figure S2: qRT-PCR_BPIV3; Figure S3: qRT-PCR_BVDV; Figure S4: qRT-PCR_BRSV; Table S1: Primers and probes sequences; Table S2: Row data results.

Author Contributions: Conceptualization, V.M. and F.S.; methodology, D.G. and S.R.; formal analysis, S.Š. and M.Đ.; investigation, M.N. and B.M.; resources, V.M.; writing—original draft preparation, M.D.; writing—review and editing, D.G.; supervision, V.M. All authors have read and agreed to the published version of the manuscript.

Funding: The study was funded by the Serbian Ministry of Science, Technological Development, and Innovation (Contract No 451-03-66/2024-03/200030).

Institutional Review Board Statement: Not applicable.

Informed Consent Statement: Not applicable.

Data Availability Statement: The data underlying this study are available upon request. Interested parties may contact the corresponding author for access to the dataset, subject to any confidentiality or data sharing agreements.

Conflicts of Interest: Snežana Ristevski works for Delta Vet Med doo and Filip Spasojević works for Al Dahra Srbija doo. There are no relevant financial or non-financial competing interests to report, and the authors declare no conflicts of interests.

References

1. Bell, R.L.; Turkington, H.L.; Cosby, S.L. The Bacterial and Viral Agents of BRDC: Immune Evasion and Vaccine Developments. *Vaccines* **2021**, *9*, 337. [CrossRef]
2. Johnson, K.K.; Pendell, D.L. Market Impacts of Reducing the Prevalence of Bovine Respiratory Disease in United States Beef Cattle Feedlots. *Front. Vet. Sci.* **2017**, *4*, 189. [CrossRef]
3. Loneragan, G.H.; Dargatz, D.A.; Morley, P.S.; Smith, M.A. Trends in Mortality Ratios among Cattle in US Feedlots. *J. Am. Vet. Med. Assoc.* **2001**, *219*, 1122–1127. [CrossRef]
4. Lemon, J.L.; McMenamy, M.J. A Review of UK-Registered and Candidate Vaccines for Bovine Respiratory Disease. *Vaccines* **2021**, *9*, 1403. [CrossRef]
5. Werid, G.M.; Miller, D.; Hemmatzadeh, F.; Messele, Y.E.; Petrovski, K. An Overview of the Detection of Bovine Respiratory Disease Complex Pathogens Using Immunohistochemistry: Emerging Trends and Opportunities. *J. Vet. Diagn. Investig.* **2024**, *36*, 12–23. [CrossRef]
6. Hause, B.M.; Ducatez, M.; Collin, E.A.; Ran, Z.; Liu, R.; Sheng, Z.; Armien, A.; Kaplan, B.; Chakravarty, S.; Hoppe, A.D.; et al. Isolation of a Novel Swine Influenza Virus from Oklahoma in 2011 Which Is Distantly Related to Human Influenza C Viruses. *PLoS Pathog.* **2013**, *9*, e1003176. [CrossRef]
7. Brito, B.P.; Frost, M.J.; Anantanawat, K.; Jaya, F.; Batterham, T.; Djordjevic, S.P.; Chang, W.S.; Holmes, E.C.; Darling, A.E.; Kirkland, P.D. Expanding the Range of the Respiratory Infectome in Australian Feedlot Cattle with and without Respiratory Disease Using Metatranscriptomics. *Microbiome* **2023**, *11*, 158. [CrossRef]

8. Nissly, R.H.; Zaman, N.; Ibrahim, P.A.S.; McDaniel, K.; Lim, L.; Kiser, J.N.; Bird, I.; Chothe, S.K.; Bhushan, G.L.; Vandegrift, K.; et al. Influenza C and D Viral Load in Cattle Correlates with Bovine Respiratory Disease (BRD): Emerging Role of Orthomyxoviruses in the Pathogenesis of BRD. *Virology* **2020**, *551*, 10–15. [CrossRef]
9. Kurcubic, V.; Djokovic, R.; Ilic, Z.; Vaskovic, N.; Petrovic, M. Bovine Respiratory Disease Complex (BRDC): A Review of Lung Lesions and Reducing of Quality of Carcasses. *Biotechnol. Anim. Husb.* **2019**, *35*, 209–217. [CrossRef]
10. Kamel, M.S.; Davidson, J.L.; Verma, M.S. Strategies for Bovine Respiratory Disease (BRD) Diagnosis and Prognosis: A Comprehensive Overview. *Animals* **2024**, *14*, 627. [CrossRef]
11. Cavirani, S. Immunization of Calves and Herd Immunity to Bovine Respiratory Disease Complex (BRDC). *Large Anim. Rev.* **2019**, *25*, 17–24.
12. Hoffmann, B.; Depner, K.; Schirrmeier, H.; Beer, M. A Universal Heterologous Internal Control System for Duplex Real-Time RT-PCR Assays Used in a Detection System for Pestiviruses. *J. Virol. Methods* **2006**, *136*, 200–209. [CrossRef]
13. Kishimoto, M.; Tsuchiaka, S.; Rahpaya, S.S.; Hasebe, A.; Otsu, K.; Sugimura, S.; Kobayashi, S.; Komatsu, N.; Nagai, M.; Omatsu, T.; et al. Development of a One-Run Real-Time PCR Detection System for Pathogens Associated with Bovine Respiratory Disease Complex. *J. Vet. Med. Sci.* **2017**, *79*, 517–523. [CrossRef]
14. Abril, C.; Engels, M.; Liman, A.; Hilbe, M.; Albini, S.; Franchini, M.; Suter, M.; Ackermann, M. Both Viral and Host Factors Contribute to Neurovirulence of Bovine Herpesviruses 1 and 5 in Interferon Receptor-Deficient Mice. *J. Virol.* **2004**, *78*, 3644–3653. [CrossRef] [PubMed]
15. Henritzi, D.; Hoffmann, B.; Wacheck, S.; Pesch, S.; Herrler, G.; Beer, M.; Harder, T.C. A Newly Developed Tetraplex Real-Time RT-PCR for Simultaneous Screening of Influenza Virus Types A, B, C and D. *Influenza Other Respir. Viruses* **2019**, *13*, 71–82. [CrossRef]
16. Thonur, L.; Maley, M.; Gilray, J.; Crook, T.; Laming, E.; Turnbull, D.; Nath, M.; Willoughby, K. One-Step Multiplex Real Time RT-PCR for the Detection of Bovine Respiratory Syncytial Virus, Bovine Herpesvirus 1 and Bovine Parainfluenza Virus 3. *BMC Vet. Res.* **2012**, *8*, 37. [CrossRef]
17. Kwasnik, M.; Rola, J.; Rozek, W. Influenza D in Domestic and Wild Animals. *Viruses* **2023**, *15*, 2433. [CrossRef]
18. Gaudino, M.; Moreno, A.; Snoeck, C.J.; Zohari, S.; Saegerman, C.; O'donovan, T.; Ryan, E.; Zanni, I.; Foni, E.; Sausy, A.; et al. Title: Emerging Influenza D Virus Infection in European Livestock as Determined in Serology Studies: Are We Underestimating Its Spread over the Continent? *Transbound. Emerg. Dis.* **2021**, *68*, 1125–1135. [CrossRef] [PubMed]
19. Sapundžić, Z.Z.; Ninković, M.; Glišić, D.; Milovanović, B.; Žutić, J.; Kureljušić, B.; Milićević, V. Seroprevalence of Bovine Viral Diarrhea on Smallholder Farms on the Territory of Belgrade City. *Acta Vet. Belgrade* **2021**, *71*, 318–325. [CrossRef]
20. Hashemi, M.; Bakhshesh, M.; Khezri, M.; Gharagouzlouian, M.M.; Tavakoli, G. A Two-Year Serological Study of Bovine Viral Diarrhea Virus, Bovine Alphaherpesvirus 1 and Bovine Parainfluenza Virus Type 3 in Qazvin Dairy Cattle Farms, Northwestern of Iran. *Vet. Arh.* **2022**, *92*, 1–10. [CrossRef]
21. Veljović, L.; Knežević, A.; Milić, N.; Krnjaić, D.; Miković, R.; Zorić, A.; Marković, M.; Milicević, V.; Stamenković, M.; Stanojević, M.; et al. Isolation and Molecular Detection of Bovine Parainfluenza Virus Type 3 in Cattle in Serbia. *Acta Vet. Belgrade* **2016**, *66*, 509–519. [CrossRef]
22. Werid, G.M.; Van, T.D.; Miller, D.; Hemmatzadeh, F.; Fulton, R.W.; Kirkwood, R.; Petrovski, K. Bovine Parainfluenza-3 Virus Detection Methods and Prevalence in Cattle: A Systematic Review and Meta-Analysis. *Animals* **2024**, *14*, 494. [CrossRef]
23. Makoschey, B.; Berge, A.C. Review on Bovine Respiratory Syncytial Virus and Bovine Parainfluenza—Usual Suspects in Bovine Respiratory Disease—A Narrative Review. *BMC Vet. Res.* **2021**, *17*, 261. [CrossRef]
24. Gaudino, M.; Nagamine, B.; Ducatez, M.F.; Meyer, G. Understanding the Mechanisms of Viral and Bacterial Coinfections in Bovine Respiratory Disease: A Comprehensive Literature Review of Experimental Evidence. *Vet. Res.* **2022**, *53*, 70. [CrossRef]
25. Ohlson, A.; Alenius, S.; Tråvén, M.; Emanuelson, U. A Longitudinal Study of the Dynamics of Bovine Corona Virus and Respiratory Syncytial Virus Infections in Dairy Herds. *Vet. J.* **2013**, *197*, 395–400. [CrossRef]
26. Bidokhti, M.R.M.; Tråvén, M.; Ohlson, A.; Zarnegar, B.; Baule, C.; Belák, S.; Alenius, S.; Liu, L. Phylogenetic Analysis of Bovine Respiratory Syncytial Viruses from Recent Outbreaks in Feedlot and Dairy Cattle Herds. *Arch. Virol.* **2012**, *157*, 601–607. [CrossRef]
27. Hoppe, I.B.A.L.; de Medeiros, A.S.R.; Arns, C.W.; Samara, S.I. Bovine Respiratory Syncytial Virus Seroprevalence and Risk Factors in Non-Vaccinated Dairy Cattle Herds in Brazil. *BMC Vet. Res.* **2018**, *14*, 208. [CrossRef]
28. Lachowicz-Wolak, A.; Klimowicz-Bodys, M.D.; Płoneczka-Janeczko, K.; Bykowy, M.; Siedlecka, M.; Cinciała, J.; Rypuła, K. The Prevalence, Coexistence, and Correlations between Seven Pathogens Detected by a PCR Method from South-Western Poland Dairy Cattle Suffering from Bovine Respiratory Disease. *Microorganisms* **2022**, *10*, 1487. [CrossRef]
29. Klem, T.B.; Sjurseth, S.K.; Sviland, S.; Gjerset, B.; Myrmel, M.; Stokstad, M. Bovine Respiratory Syncytial Virus in Experimentally Exposed and Rechallenged Calves; Viral Shedding Related to Clinical Signs and the Potential for Transmission. *BMC Vet. Res.* **2019**, *15*, 156. [CrossRef]
30. Brock, J.; Lange, M.; Guelbenzu-Gonzalo, M.; Meunier, N.; Vaz, A.M.; Tratalos, J.A.; Dittrich, P.; Gunn, M.; More, S.J.; Graham, D.; et al. Epidemiology of Age-Dependent Prevalence of Bovine Herpes Virus Type 1 (BoHV-1) in Dairy Herds with and without Vaccination. *Vet. Res.* **2020**, *51*, 124. [CrossRef]
31. Barrett, D.; Parr, M.; Fagan, J.; Johnson, A.; Tratalos, J.; Lively, F.; Diskin, M.; Kenny, D. Prevalence of Bovine Viral Diarrhoea Virus (BVDV), Bovine Herpes Virus 1 (BHV 1), Leptospirosis and Neosporosis, and Associated Risk Factors in 161 Irish Beef Herds. *BMC Vet. Res.* **2018**, *14*, 8. [CrossRef]

32. Raaperi, K.; Nurmoja, I.; Orro, T.; Viltrop, A. Seroepidemiology of Bovine Herpesvirus 1 (BHV1) Infection among Estonian Dairy Herds and Risk Factors for the Spread within Herds. *Prev. Vet. Med.* **2010**, *96*, 74–81. [CrossRef]
33. Woodbine, K.A.; Medley, G.F.; Moore, S.J.; Ramirez-Villaescusa, A.M.; Mason, S.; Green, L.E. A Four Year Longitudinal Sero-Epidemiological Study of Bovine Herpesvirus Type-1 (BHV-1) in Adult Cattle in 107 Unvaccinated Herds in South West England. *BMC Vet. Res.* **2009**, *5*, 5. [CrossRef]
34. Paller, T.; Hostnik, P.; Pogacnik, M.; Toplak, I. The Prevalence of Ten Pathogens Detected by a Real-Time PCR Method in Nasal Swab Samples Collected from Live Cattle with Respiratory Disease. *Slov. Vet. Res.* **2017**, *54*, 101–107.
35. Wernicki, A.; Urban-Chmiel, R.; Rola, J.; Socha, W.; Stęgierska, D.; Dec, M.; Puchalski, A. Prevalence of Bovine Herpes Virus Type 1 in Small Herds of Young Beef Cattle in South-Eastern Poland-a Preliminary Study. *Bull. Vet. Inst. Pulawy* **2015**, *59*, 467–471. [CrossRef]
36. Su, N.; Wang, Q.; Liu, H.Y.; Li, L.M.; Tian, T.; Yin, J.Y.; Zheng, W.; Ma, Q.X.; Wang, T.T.; Li, T.; et al. Prevalence of Bovine Viral Diarrhea Virus in Cattle between 2010 and 2021: A Global Systematic Review and Meta-Analysis. *Front. Vet. Sci.* **2023**, *9*, 1086180. [CrossRef]
37. Petrović, T.; Lazić, S.; Lupulović, D.; Bugarski, D.; Đuričin, B. *Rezultati Ispitivanja Raširenosti BVDV Infekcije u Velikim i Malim Zapatima Goveda Na Području Južnobačkog i Sremskog Okruga*; Scientific Veterinary Institute "Novi Sad": Novi Sad, Serbia, 2008; Volume 1.
38. Mehinagic, K.; Pilo, P.; Vidondo, B.; Stokar-Regenscheit, N. Coinfection of Swiss Cattle with Bovine Parainfluenza Virus 3 and Mycoplasma Bovis at Acute and Chronic Stages of Bovine Respiratory Disease Complex. *J. Vet. Diagn. Investig.* **2019**, *31*, 674–680. [CrossRef]

animals

MDPI

Article

Determining the Economically Optimum Metaphylactic Strategy for Cattle Cohorts of Varied Demographic Characteristics

Dannell J. Kopp [1], Robert L. Larson [1], Phillip A. Lancaster [1], Bradley J. White [1,*], Kristen J. Smith [2] and Dustin L. Pendell [3]

[1] Beef Cattle Institute, Department of Clinical Sciences, College of Veterinary Medicine, Kansas State University, Manhattan, KS 66506, USA
[2] Private Consultant, Baldwin City, KS 66006, USA
[3] Department of Agricultural Economics, College of Agriculture, Kansas State University, Manhattan, KS 66506, USA
* Correspondence: bwhite@vet.k-state.edu

Simple Summary: The administration of antibiotics to a whole cohort of high-risk feedlot cattle upon arrival to the feedlot is a common practice used by producers to control respiratory disease. A model was developed that utilizes an economic endpoint to determine the best application of antibiotics based on demographics of the cattle typically included in feedlot records. Determining the best situations to use and not use antibiotics helps to not only decrease cases of respiratory disease but also to ensure continued usefulness of antimicrobials. The goal of this study is to identify key characteristics of cohorts of cattle that did not receive metaphylaxis but that would benefit economically from the use of arrival antibiotics.

Abstract: Metaphylactic antibiotic use in feeder cattle is a common practice to control respiratory disease. Antimicrobial stewardship is important to ensure continued efficacy and to protect animal welfare. The objective of this study is to identify characteristics of cohorts of cattle that had not received metaphylaxis that would have benefited economically from the use of metaphylaxis. Cohorts (n = 12,785; 2,206,338 head) from 13 feedlots that did not receive metaphylaxis were modeled using an economic model to estimate net returns for three metaphylactic options. Logistic regression models with covariates for entry weight, sex, average daily weight gain, number of animals per cohort, and days on feed, with feedlot as a random effect, were used to determine the model-adjusted probability of cohorts benefiting economically from metaphylaxis. Most (72%) cohorts in this data set that had not received metaphylaxis at arrival would not economically benefit from metaphylaxis. Sex, entry weight category, number of cattle in the cohort, and average daily weight gain were associated with the likelihood of benefitting economically from metaphylaxis. The results illustrated that cattle cohort demographics influenced the probability that cohorts would benefit economically from metaphylaxis and the type of metaphylaxis utilized, and integrating this information has the potential to influence the metaphylaxis decision.

Keywords: bovine respiratory disease; economics; metaphylactic antibiotics

1. Introduction

The bovine respiratory disease (BRD) complex is a multi-faceted disease involving bacterial and viral agents, environmental factors, commingling, transport, and other stressors [1–5]. In 2010, cattle death losses in the United States due to respiratory problems equated to $643,146,000 [6]. While mortality due to BRD may be a major economic loss, BRD morbidity is also important when including costs of treatment, labor, reduced average daily weight gain (ADG), and feed conversion [7–11]. Negative effects on ADG result in smaller, lower-marbling carcasses, with losses ranging from $23.23 to $54.01 per carcass for cattle treated for BRD [12]. Cattle treated for BRD more than once also result in lower carcass grades compared to those that were only treated once or not at all [13]. Combining all sources of potential loss, BRD leads to substantial economic losses for cattle producers. Decreased prevalence of BRD within the livestock industry can also lead to varied effects on many sectors of the industry, including potentially lowering beef prices for the consumer and benefiting society [14]. A true evaluation of BRD cost should include mortality and all the expenses associated with morbidity [15].

Antimicrobial metaphylaxis has been implemented in an effort to curb economic losses, as well as improve the animal welfare of cattle entering feedlots [16]. Metaphylaxis is the use of a U.S. Food and Drug Administration (FDA)-approved antimicrobial agent to treat an entire cohort of cattle that are at high risk for a disease process that would be modified by treatment at or near the time of arrival to a feedlot [17]. Decreased risk of BRD-related morbidity and mortality are commonly seen results of implementing metaphylaxis [18–21].

The impact of BRD is highly variable among cohorts of cattle [22,23]. In addition, as an indication of metaphylactic efficacy, odds ratios for BRD morbidity cumulative incidence and BRD mortality cumulative incidence vary greatly between different comparisons of common metaphylactic drugs [24]. The combination of inconsistent disease impact and variable metaphylactic efficacy makes it difficult to predict which cohorts will benefit from metaphylactic treatment for the control of BRD.

Currently, metaphylaxis decisions are made based on risk classification of the cohort [17,19]. Risk classification assigned at the feedlot is made based on both subjective and objective measures. While much work has been done to show that metaphylaxis can reduce morbidity and mortality, because of differences in efficacy and cost between metaphylactic options, optimum metaphylaxis decisions must be based on more information than risk classification alone [25,26]. Looking at the decision strictly from an economic outcome helps remove some subjectivity, and incorporating metaphylactic efficacy and cost brings additional important factors to the decision-making process.

The objective of this study is to identify characteristics of cohorts of cattle that had not received metaphylaxis that would have benefited economically from the use of metaphylaxis. Specifically, three metaphylactic options are evaluated: no metaphylaxis (NOMET), low-cost/low-efficacy metaphylaxis (LCLE), and high-cost/high-efficacy metaphylaxis (HCHE), to discover information to update antimicrobial use decisions leading to improved antimicrobial stewardship and economic success. The development of an economic model to determine the differences in net returns between the three metaphylactic options will be used to identify cohort demographics associated with each optimized metaphylactic option.

2. Materials and Methods

2.1. Ethical Statement

Institutional Animal Care and Use Committee (IACUC) approval was not required as historical operational data were used for the analysis and no procedures were performed on cattle specifically for this research.

2.2. Study Design

Retrospective, observational, cohort-level data from 13 feedlots over a 5-year period (2016–2020) were collected under confidentiality agreements with feedlot collaborators. Data were combined with economic and statistical models to assess the potential economic impact of metaphylaxis in cohorts that did not receive metaphylaxis. A cohort was defined as a purchased group of cattle that entered and exited the feedlot together. BRD morbidity was defined as cattle identified with BRD and treated with antimicrobials. BRD mortality was defined as cattle that died following treatment for BRD or that were classified as dying of BRD even if not previously treated (i.e., pen deads).

2.3. Exclusion Criteria

Exclusion criteria were applied to the feedlot data to remove potential data entry errors and to guarantee that only cohorts with complete data were included. For this analysis, only cohorts that did not receive metaphylaxis upon arrival were included. The average entry weight of cohorts had to be greater than 181 kg and less than 499 kg. Cohorts of mixed sex were excluded, in order to be able to analyze differences in economic benefits between cohorts of steers and cohorts of heifers. Data were filtered to only include records where cohort size was greater than 40 cattle at arrival. Total cohort days on feed was limited to less than 400 days and greater than 70 days. Cohort BRD morbidity risk was required to be greater than or equal to BRD mortality risk, meaning the case fatality risk of the cohorts could not be greater than 1, which would have allowed for more BRD deaths than calves treated for BRD.

2.4. Study Variables from Cohort Records

The following data were acquired from cohort records: number of head purchased, whole-cohort entry weight, number treated for BRD, number of deaths attributed to BRD, days on feed, number of head sold, whole-cohort exit weight, and pounds of feed consumed. The subsequent adjusted and calculated variables were computed in order to have a consistent comparison between metaphylactic options. All cohorts were adjusted to a constant entry date of 5 October 2018, with exit dates based on actual days on feed (constant entry date allowed for the comparison of cohort demographic effects on the economic outcomes while holding cattle and feed prices steady). The whole-cohort entry weight was divided by number of head purchased to determine the average entry weight per head.

The average daily weight gain of cattle within a cohort was calculated as the average entry weight per head subtracted from the average exit weight per head divided by days on feed to compute deads-out ADG. Bovine respiratory disease morbidity and mortality risk was provided in the original data from collaborating feedlots. Morbidity attributed to other causes beyond BRD was not included in the model. The case fatality risk as a depiction of treatment success was calculated as number of deaths from BRD divided by number of cattle treated for BRD within each cohort. Deaths not attributed to BRD were designated as "other deaths" and were assumed to be equal proportions regardless of metaphylactic options.

Head days for each cohort were calculated by assuming that BRD deaths occurred at day 50, per expert opinion. Other deaths were assumed to occur at the midpoint for days on feed for that cohort, and cattle that finished the feeding phase had head days equivalent to days on feed. Head days were used to calculate expenses accrued on a daily time-step by the number of cattle in the cohort present at each day, such as yardage and feed cost.

Arrival processing and yardage costs were acquired for each cohort from the cooperating feedlot. The whole-cohort arrival processing cost was divided by the number of cattle upon feedlot arrival and then averaged across all cohorts to determine an average processing cost per head across all yards. The processing cost was assigned as \$23.60 per head upon feedlot entry and was based on data provided by cooperating feedlots. The whole-cohort yardage cost was divided by head days and then averaged across all cohorts to determine an average yardage cost per head day across all yards. A yardage cost of \$0.35 per head day was used in the model.

2.5. Study Variables—Cattle Prices

Feeder cattle purchase prices for 5 October 2018 (Table 1) [27–31] were acquired from an online database. A purchase cost was then calculated based on the average entry weight for each cohort and the purchase price relative to entry weight category. Live cattle sale prices were the weekly prices reported from November 2018 to October 2019 (Table 2) [32]. The sale price was assigned based on the cohort exit date from the feedlot, with the exit date being equivalent to 5 October 2018, plus the actual days on feed of the cohort. The sale price and the cohort exit weights were used to determine the income for each original cohort and its two metaphylaxis treatment-generated cohorts.

Table 1. Feeder cattle purchase prices as of 5 October 2018 [27–31].

Entry Weight Category	Price (USD/kg)	Price (USD/cwt)
181.4–226.8 kg:	4.11	186.36
226.8–272.2 kg:	3.83	173.65
272.2–317.5 kg:	3.60	163.28
317.5–362.9 kg:	3.50	158.73
362.9–499.0 kg:	3.39	153.67

Table 2. Fed cattle prices, monthly range of weekly sale prices [32].

Sale Month	Price (USD/kg)	Price (USD/cwt)
November 2018:	2.5–2.59	113.53–117.54
December 2018:	2.61–2.7	118.40–122.65
January 2019:	2.7–2.74	122.63–124.06
February 2019:	2.73–2.78	123.73–126.15
March 2019:	2.77–2.83	125.83–128.50
April 2019:	2.76–2.82	125.19–127.75
May 2019:	2.54–2.92	115.28–132.27
June 2019:	2.43–2.5	110.29–113.48
July 2019:	2.43–2.5	110.41–113.37
August 2019:	2.31–2.49	104.89–112.78
September 2019:	2.2–2.31	99.86–104.61
October 2019:	2.34–2.43	106.26–110.03

2.6. Study Variables—Metaphylactic Treatment Prices

The pricing of metaphylactic options was based on various antibiotic agents at bulk price from an online veterinary retailer as of June 2022, in combination with expert opinion [33]. The costs for the LCLE and HCHE metaphylactic options were set at $0.02 per kg and $0.08 per kg of entry weight, respectively. The cost of metaphylaxis was in addition to the processing costs.

2.7. Study Variables—Feed Costs

The feed ration price per kg was determined using historical commodity and feedstuffs prices. Kansas State University's Focus on Feedlots Monthly Report was used for alfalfa hay and corn prices [34]. Agriculture Marketing Resource Center's Weekly Ethanol, Distillers Grain, and Corn Prices was used to determine the price of wet distillers grains based on an average of the monthly prices from Illinois, Iowa, Nebraska, South Dakota, and Wisconsin [35]. A feed ration consisting of 10% alfalfa hay, 60% corn, and 30% wet distillers grains was developed. Prices for alfalfa, corn, and wet distillers grains were determined to be $170.24 per ton ($0.19 per kg), $137.50 per ton ($0.15 per kg), and $46.12 per ton ($0.05 per kg), respectively. Using these values, feed was priced at $0.18 per kg of feed. For NOMET, the total pounds of feed consumed was acquired from data records and used to determine feed costs. When cohort data were modified for LCLE and HCHE metaphylaxis treatment-generated cohorts, the total pounds of feed consumed was modified by a proportion of change in head days to account for increased head days within the cohort due to fewer BRD deaths.

2.8. Study Variables Based on Differences in Metaphylactic Efficacy on Odds Ratios for Cumulative BRD Morbidity

A systematic review and mixed treatment meta-analysis of metaphylaxis was used to identify that there are differences in efficacy among metaphylactic antimicrobials which allowed us to create higher-efficacy and lower-efficacy categories [24]. These higher-

and lower-efficacy categories were not intended to indicate specific antimicrobials but were used to inform the metaphylactic efficacy odds ratios of 0.6 and 0.3 for LCLE and HCHE, respectively.

Each cohort was modeled three times: first, the actual performance of the cohort (i.e., no-metaphylaxis NOMET); second, the modeled performance of the cohort if it had received LCLE; and third, the modeled performance of the cohort if it had received HCHE. The metaphylactic efficacy odds ratio was used to modify the morbidity risk of each NOMET cohort to determine a new, reduced morbidity risk within the cohort when either LCLE or HCHE metaphylaxis was modeled. The new morbidity risk was used to determine the number of animals within each metaphylaxis treatment-generated cohort that was treated for BRD. The original case fatality risk of each baseline cohort (NOMET) was then used to determine the number of animals that died from BRD for each metaphylaxis treatment-generated comparison cohort. Other deaths were assumed to not be differentially affected by the use of a metaphylactic antibiotic. A lower morbidity risk and subsequently lower mortality risk increased both income and costs for LCLE- and HCHE-modeled cohorts compared to the income and costs modeled for the associated NOMET cohort.

Using each NOMET cohort's ADG we calculated a healthy-adjusted ADG, as well as a BRD-treated-adjusted ADG, in relation to the number of morbid calves in the NOMET cohort. These were calculated on the basis of a 5% decrease in weight gain of the BRD-treated calves [12]. These adjusted ADG values were used to determine the group-level ADG for each metaphylaxis treatment-generated cohort based on the adjusted number of healthy and BRD-morbid cattle for both LCLE and HCHE options. The group-level ADG was then used to calculate the total exit weight for each metaphylaxis treatment-generated cohort.

The number of head days for the LCLE and HCHE cohort models were based on the metaphylaxis treatment-adjusted number of cattle that died from BRD, relative to the designated morbidity and mortality reduction assigned for the LCLE and HCHE, and the number of other deaths provided in the NOMET cohort. Bovine respiratory disease deaths were assumed to die on day 50 while other deaths were assumed to die at the midpoint of days on feed.

Morbid cattle that were pulled and treated for BRD were charged for the high-cost antibiotic priced at $0.08 per kg as well as being charged a $1.50 chute fee per head. For dosing purposes, the weight of morbid cattle was based on the average cohort entry weight into the feedlot.

2.9. Model Comparison of Net Return between Metaphylactic Options

A net return economic model using enterprise budgets was developed to assess the optimal metaphylactic option of cohorts of U.S. feedlot cattle that had not received metaphylaxis at arrival based on the highest net return among the three metaphylactic options: NOMET, LCLE, and HCHE. For analysis, a logistic regression model was used to initially compare NOMET to a "yes" metaphylaxis variable (YESMET) which comprised both LCLE and HCHE, followed by a second logistic regression model to analyze LCLE versus HCHE within the YESMET-modeled cohorts.

Net returns were calculated as the difference between total revenue and total expenses. Total expenses included the purchase cost of feeder cattle, processing cost, yardage cost, BRD treatment cost, chute fee cost, and feed cost, as well as the cost of metaphylaxis. Net returns for all cohorts were compared across NOMET, LCLE, and HCHE, and the optimal decision for each cohort was selected based on the highest net return.

Economic calculations and metaphylaxis adjustments were completed using Excel (Microsoft Excel, Microsoft Office Professional Plus 2019, Microsoft Corporation, Redmond, WA, USA). Data analysis was completed using RStudio (R Core Team, 2022). The association of sex, entry weight, cohort size, exit weight, ADG, and days on feed with the probability of benefiting economically from metaphylaxis options was evaluated with a logistic regression model with feedlot as a random effect using the glm() function. Data variables were evaluated for interactions and none were found. Two logistic regression models were completed with the first model comparing NOMET to a "yes" metaphylaxis variable that combined LCLE and HCHE into one outcome, and the second model comparing LCLE to HCHE within the cohorts that were YESMET in the first regression analysis. Among cohorts benefiting economically from metaphylaxis, the association of the same variables with the probability of benefiting from HCHE metaphylaxis versus LCLE metaphylaxis was evaluated with a logistic regression model with feedlot as a random effect.

3. Results

Records for 16,809 cohorts were originally acquired from collaborators. The application of the exclusion criteria outlined above resulted in 12,785 cohorts of feedlot cattle in the dataset. The final dataset of 12,785 cohorts encompassed 4038 cohorts of heifers and 8747 cohorts of steers. These cohorts included 2,206,338 head of cattle that originated from 38 states across the United States; 631,521 head were heifers and 1,574,817 head were steers. The analysis of the included cohorts found the median percent morbidity risk was 5.1% with an interquartile range of 2.5% to 10% (Figure 1).

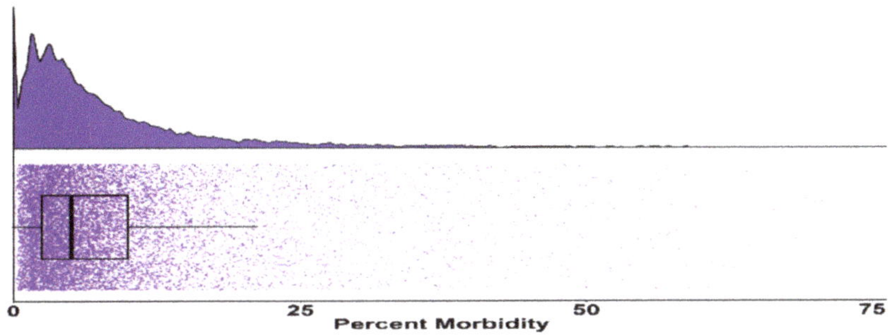

Figure 1. Distribution of cohort morbidity risk displayed in a raincloud plot. The upper section shows density of morbidity percentage due to bovine respiratory disease for each cohort. The lower portion shows a scatter plot of each cohort by morbidity percentage with a box-and-whisker plot overlay.

Metaphylaxis adjustments (LCLE and HCHE) resulted in a greater sale weight and less BRD treatment costs, as well as an increased arrival processing cost, increased feed cost due to increased feed consumed, increased yardage, and other costs associated with less disease and death compared to NOMET. The optimal metaphylactic choice based on the economic model output varied among cohorts with a distribution of cohorts falling into NOMET (72.0%; 9204/12,785), LCLE (25.4%; 3242/12,785), and HCHE (2.7%; 339/12,785). Statistical analysis included two logistic regressions that first compared NOMET (72%) to YESMET (28%) and then compared LCLE to HCHE within the YESMET cohorts.

When comparing NOMET to YESMET, sex, the entry weight category, number of cattle in the cohort, and the average daily weight gain category were found to be statistically significant. Cohorts of steers had a higher probability (31.2 ± 3.3%) of benefiting economically from metaphylaxis in comparison to cohorts of heifers (19.9 ± 2.5%) ($p < 0.05$). Cohorts within the three lowest average entry weight intervals (37 kg intervals) had a higher probability of benefiting economically from metaphylaxis than the other four weight categories (Figure 2). Cohorts of less than 100 head of cattle had a higher probability (30.0 ± 3.3%) of benefiting economically from metaphylaxis in comparison to cohorts with 100 to 200 head of cattle (25.1 ± 2.9%) ($p < 0.05$) and those cohorts with over 200 head of cattle (20.8 ± 2.6%) ($p < 0.05$). Cohorts with an ADG of 0.45–1.36 kg/day had a higher probability (8.3 ± 0.5%) of benefiting economically from metaphylaxis compared to those with ADG greater than 1.36 kg/day (Figure 3).

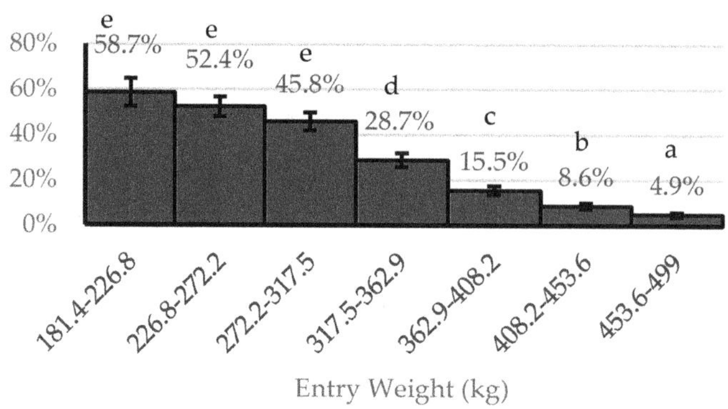

Figure 2. Probability of benefiting economically from metaphylaxis (YESMET) by entry weight. [abcde] Bars without a common superscript letter differ at $p \leq 0.05$.

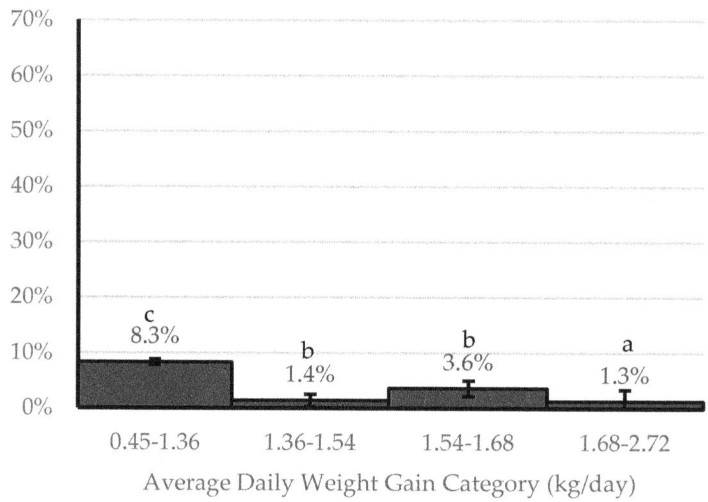

Figure 3. Probability of benefiting economically from metaphylaxis (YESMET) by average daily weight gain categories. [abc] Bars without a common superscript letter differ at $p \leq 0.05$.

Within the YESMET cohorts, the entry weight category, number of cattle in the cohort, and average daily weight gain were associated with the probability of HCHE versus LCLE metaphylaxis having a higher net return. Of cohorts benefiting economically from metaphylaxis, those within the three lowest entry weight categories had the highest probability of benefiting economically from HCHE metaphylaxis (Figure 4). Of cohorts that benefited from metaphylaxis, cohorts of less than 100 head had a higher probability (5.3 $\pm$ 1.4%) ($p < 0.05$) of benefiting from HCHE metaphylaxis in comparison to cohorts of 100 to 200 head of cattle (3.3 $\pm$ 0.9%) and cohorts of over 200 head of cattle (3.2 $\pm$ 0.9%). Of those that benefited from metaphylaxis, cohorts with an ADG of 0.45–1.36 kg/day had the highest probability of benefiting economically from HCHE metaphylaxis and cohorts with an ADG of 1.36–1.54 kg/day and 1.54–1.68 kg/day had a higher probability of benefiting economically from HCHE metaphylaxis compared to cohorts with an ADG of 1.68–2.72 kg/day (Figure 5).

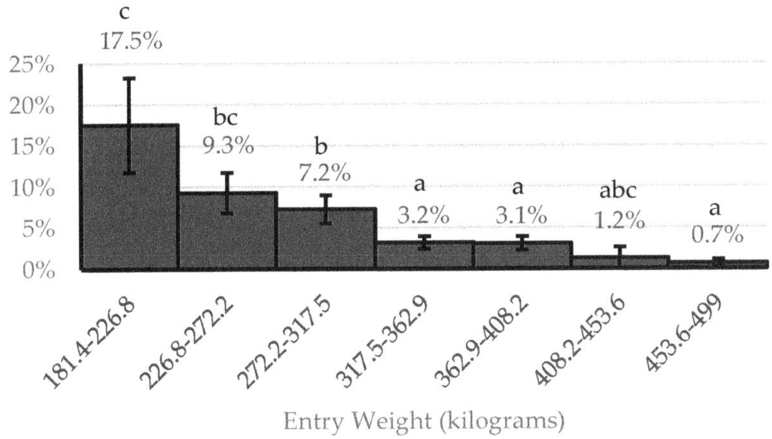

Figure 4. Among cohorts benefiting economically from metaphylaxis (YESMET), the probability of benefiting economically from HCHE by entry weight. [abc] Bars without a common superscript letter differ at $p \leq 0.05$.

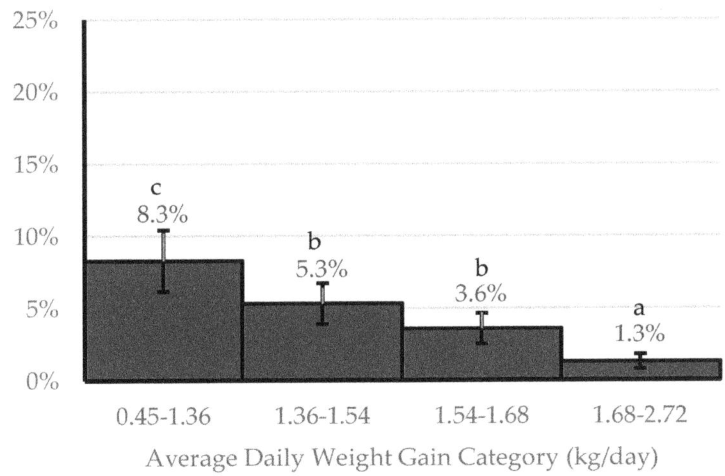

Figure 5. Among cohorts benefiting economically from metaphylaxis (YESMET), the probability of benefiting economically from HCHE by average daily weight gain categories. [abc] Bars without a common superscript letter differ at $p \leq 0.05$.

4. Discussion

Modeling the administration of metaphylaxis changed both income and costs resulting in different net returns for the three modeled scenarios for each original cohort. When comparing cohorts of feedlot cattle that had not received metaphylaxis across NOMET, LCLE, and HCHE, most (72%) cohorts in this data set would not economically benefit from metaphylaxis. Our results differ from Dennis et al. [26] who modeled net return distributions for metaphylactic decisions for cattle at high risk for BRD. Because the data set used in our study only included cohorts that did not receive metaphylaxis, we assume that the perceived risk of BRD in our cohorts was lower than for cohorts modeled by Dennis et al. [26].

Based on the cohorts used in this analysis, we found a median cohort percent morbidity of 5.1%. Prior research showed a 12.8% bovine respiratory disease morbidity risk in feeder cattle via surveys of feedlots in Kansas, Nebraska, Texas, North Dakota, and other states, which differs substantially from our results [36]. Because we excluded cohorts that received metaphylaxis at arrival, it is not surprising that morbidity was lower in our study. Cattle chosen to receive metaphylaxis at feedlots typically have an increased risk of developing disease, so exclusion of these animals would lead to decreased morbidity within this data set [17].

Steers had a greater probability of benefiting economically from metaphylaxis compared to heifers. Other studies have estimated that steers have a mean ADG of 1.62 kg per day while heifers have a mean ADG of 1.36 kg per day [37]. With this in mind, a 5% decrease in ADG seen in cattle affected by BRD would have a greater impact on cohorts of steers because of their greater ADG. Metaphylaxis thus provides more improvement in sale weight within steer cohorts over heifer cohorts.

Upon arrival to the feedlot, lightweight cohorts of cattle are considered to be at greater risk for BRD in comparison to heavier cattle [17]. The same BRD odds ratio benefit from metaphylaxis applied to high-morbidity-risk cohorts provides a greater reduction in number of cattle affected by BRD compared to low-morbidity-risk cohorts; therefore, metaphylaxis, and specifically, HCHE metaphylaxis, is more likely to provide economic benefits for high-risk cohorts.

Cohorts with less than 100 head of cattle were more likely to benefit economically from metaphylaxis. We do not have a hypothesis for the biological association between smaller cohort size and the economic benefit of metaphylaxis. A smaller cohort size would be associated with a proportionally higher impact of morbidity and mortality. One morbid animal in a pen of 50 head (1/50, 2.0%) would cause a significantly higher morbidity compared to a single morbid animal in a pen of 500 head (1/500, 0.2%). Another possibility for this association is that smaller cohorts of animals may be more likely to be commingled in order to fill a pen. Commingling means that cattle from different sources are put together in order to fill a pen closer to capacity. The commingling of cattle from different sources increases the likelihood of cattle developing BRD [38]. Each smaller cohort will have different levels of immunity to diseases as well as bring in different strains of bacteria and viruses, which helps to contribute to an increased likelihood of developing disease. We were surprised to see that cohorts with lower ADG were more likely to benefit from metaphylaxis and HCHE metaphylaxis.

The limitations within this study include the following: the choice of constant entry dates and pricing; set values for feed costs, chute fees, yardage costs, processing costs, and metaphylactic prices and efficacies. Constant entry dates and pricing allow for a more straight-forward comparison of net returns based solely on metaphylaxis effects. The date of 5 October 2018 was chosen to allow all cohorts to reach slaughter prior to the COVID-19 pandemic, as there were drastic changes in market values during those times. We discern that these values would be a better representation of actual figures if they were allowed to fluctuate, but the additional variation would hinder the comparison of cohort demographics across equivalent economic factors of production. Future research utilizing stochastic prices for cattle purchasing, sales, and various inputs would be useful for determining how prices affect metaphylactic decision making.

5. Conclusions

By modeling cohorts of feeder cattle for NOMET, LCLE, and HCHE metaphylaxis options, we were able to identify which strategy resulted in the greatest net return, and we were able to analyze key differences and similarities between the populations. We found that most (72%) cohorts in this data set that had not received metaphylaxis at arrival would not economically benefit from metaphylaxis, and we determined that cattle cohort demographics influenced the probability that cohorts would benefit economically from metaphylaxis and the type of metaphylaxis utilized. The results of this study as well as future published studies will provide decision makers with important information to improve the integration of multiple considerations when making metaphylaxis decisions in cohorts of cattle and management systems similar to those modeled in this study. Based on the results of this study, decision makers should recognize that NOMET was the correct decision in 72% of comparisons, and they should consider using metaphylaxis in cohorts currently not receiving the intervention when the arrival weight is low, the cohort consists of a small number of cattle, and the cohort sex is steers. In order to optimize net returns while maintaining antimicrobial stewardship and animal welfare, producers and veterinarians should continue to develop decision tools that incorporate cohort demographic factors when making metaphylaxis choices.

Author Contributions: Conceptualization, D.J.K., R.L.L., B.J.W. and D.L.P.; methodology, D.J.K. and R.L.L.; software, D.J.K. and K.J.S.; validation, D.J.K., R.L.L. and B.J.W.; formal analysis, D.J.K. and K.J.S.; investigation, D.J.K. and K.J.S.; resources, B.J.W.; data curation, K.J.S.; writing—original draft preparation, D.J.K.; writing—review and editing, R.L.L., P.A.L., B.J.W. and D.L.P.; visualization, D.J.K. and K.J.S.; supervision, R.L.L., B.J.W. and D.L.P.; project administration, B.J.W.; funding acquisition, B.J.W., P.A.L., D.L.P. and R.L.L. All authors have read and agreed to the published version of the manuscript.

Funding: This research was funded by the Foundation for Food and Agriculture Research (FFAR), grant number ICASATWG-0000000041, and scholar funding was provided by the FFAR Veterinary Student Research Fellowship Program and Kansas State University's Veterinary Research Scholars Program.

Institutional Review Board Statement: Ethical review and approval were waived for this study due to the usage of retrospective data.

Informed Consent Statement: Informed consent was obtained from all subjects involved in this study.

Data Availability Statement: Data were provided through agreements with collaborating entities and data are bound by confidentiality agreements; thus, raw data cannot be shared.

Conflicts of Interest: The authors declare no conflicts of interest. The funders had no role in the design of the study; in the collection, analyses, or interpretation of data; in the writing of the manuscript; or in the decision to publish the results.

References

1. Smith, R.A.; Step, D.L.; Woolums, A.R. Bovine Respiratory Disease: Looking Back and Looking Forward, What Do We See? *Vet. Clin. N. Am. Food Anim. Pract.* **2020**, *36*, 239–251. [CrossRef] [PubMed]
2. Martin, S.W.; Bateman, K.G.; Shewen, P.E.; Rosendal, S.; Bohac, J.E. The Frequency, Distribution and Effects of Antibodies, to Seven Putative Respiratory Pathogens, on Respiratory Disease and Weight Gain in Feedlot Calves in Ontario. *Can. J. Vet. Res. Rev. Can. Rech. Vet.* **1989**, *53*, 355–362.
3. Martin, S.W.; Bohac, J.G. The Association between Serological Titers in Infectious Bovine Rhinotracheitis Virus, Bovine Virus Diarrhea Virus, Parainfluenza-3 Virus, Respiratory Syncytial Virus and Treatment for Respiratory Disease in Ontario Feedlot Calves. *Can. J. Vet. Res. Rev. Can. Rech. Vet.* **1986**, *50*, 351–358.
4. Griffin, D.; Chengappa, M.M.; Kuszak, J.; McVey, D.S. Bacterial pathogens of the bovine respiratory disease complex. *Vet. Clin. Food Anim.* **2010**, *26*, 381–394. [CrossRef] [PubMed]
5. Taylor, J.D.; Fulton, R.W.; Lehenbauer, T.W.; Step, D.L.; Confer, A.W. The epidemiology of bovine respiratory disease: What is the evidence for predisposing factors? *Can. Vet. J.* **2010**, *51*, 1095. [PubMed] [PubMed Central]
6. National Agricultural Statistic Service. Cattle Death Loss. National Agricultural Statistics Service (NASS), Agricultural Statistics Board, United States Department of Agriculture (USDA). 2011. Available online: https://www.nass.usda.gov/Publications/Todays_Reports/reports/catlos11.pdf (accessed on 27 June 2023).
7. Smith, R.A. Impact of Disease on Feedlot Performance: A Review. *J. Anim. Sci.* **1998**, *76*, 272. [CrossRef] [PubMed]
8. Wittum, T.E.; Woollen, N.E.; Perino, L.J.; Littledike, E.T. Relationships among Treatment for Respiratory Tract Disease, Pulmonary Lesions Evident at Slaughter, and Rate of Weight Gain in Feedlot Cattle. *J. Am. Vet. Med. Assoc.* **1996**, *209*, 814–818. [CrossRef] [PubMed]
9. Thompson, P.N.; Stone, A.; Schultheiss, W.A. Use of Treatment Records and Lung Lesion Scoring to Estimate the Effect of Respiratory Disease on Growth during Early and Late Finishing Periods in South African Feedlot Cattle. *J. Anim. Sci.* **2006**, *84*, 488–498. [CrossRef] [PubMed]
10. Blakebrough-Hall, C.; McMeniman, J.P.; González, L.A. An Evaluation of the Economic Effects of Bovine Respiratory Disease on Animal Performance, Carcass Traits, and Economic Outcomes in Feedlot Cattle Defined Using Four BRD Diagnosis Methods. *J. Anim. Sci.* **2020**, *98*, skaa005. [CrossRef] [PubMed]

11. Brooks, K.R.; Raper, K.C.; Ward, C.E.; Holland, B.P.; Krehbiel, C.R.; Step, D.L. Economic Effects of Bovine Respiratory Disease on Feedlot Cattle during Backgrounding and Finishing Phases. *Prof. Anim. Sci.* **2011**, *27*, 195–203. [CrossRef]
12. Schneider, M.J.; Tait, R.G.; Busby, W.D.; Reecy, J.M. An Evaluation of Bovine Respiratory Disease Complex in Feedlot Cattle: Impact on Performance and Carcass Traits Using Treatment Records and Lung Lesion Scores. *J. Anim. Sci.* **2009**, *87*, 1821–1827. [CrossRef] [PubMed]
13. Fulton, R.W.; Cook, B.J.; Step, D.L.; Confer, A.W.; Saliki, J.T.; Payton, M.E.; Burge, L.J.; Welsh, R.D.; Blood, K.S. Evaluation of Health Status of Calves and the Impact on Feedlot Performance: Assessment of a Retained Ownership Program for Postweaning Calves. *Can. J. Vet. Res.* **2002**, *66*, 173–180. [PubMed]
14. Johnson, K.K.; Pendell, D.L. Market Impacts of Reducing the Prevalence of Bovine Respiratory Disease in United States Beef Cattle Feedlots. *Front. Vet. Sci.* **2017**, *4*, 189. [CrossRef] [PubMed]
15. White, B.J.; Larson, B.L. Impact of Bovine Respiratory Disease in U.S. Beef Cattle. *Anim. Health Res. Rev.* **2020**, *21*, 132–134. [CrossRef] [PubMed]
16. Corbin, M.J.; Gould, J.A.; Carter, B.L.; McClary, D.G.; Portillo, T.A. Effects and Economic Implications of Metaphylactic Treatment of Feeder Cattle with Two Different Dosages of Tilmicosin on the Incidence of Bovine Respiratory Disease (BRD): A Summary of Two Studies. *Bov. Pract.* **2020**, *43*, 140–152. [CrossRef]
17. Ives, S.E.; Richeson, J.T. Use of Antimicrobial Metaphylaxis for the Control of Bovine Respiratory Disease in High-Risk Cattle. *Vet. Clin. N. Am. Food Anim. Pract.* **2015**, *31*, 341–350. [CrossRef] [PubMed]
18. Daniels, T.K.; Bowman, J.G.P.; Sowell, B.F.; Branine, M.E.; Hubbert, M.E. Effects of Metaphylactic Antibiotics on Behavior of Feedlot Calves. *Prof. Anim. Sci.* **2000**, *16*, 247–253. [CrossRef]
19. Coppin, C.M.; Smock, T.M.; Helmuth, C.L.; Manahan, J.L.; Long, N.S.; Hoffman, A.A.; Carroll, J.A.; Broadway, P.R.; Burdick Sanchez, N.C.; Wells, J.E.; et al. The Effects of Administering Different Metaphylactic Antimicrobials on Growth Performance and Health Outcomes of High-Risk, Newly Received Feedlot Steers. *Transl. Anim. Sci.* **2022**, *6*, txac140. [CrossRef] [PubMed]
20. Word, A.B.; Wickersham, T.A.; Trubenbach, L.A.; Mays, G.B.; Sawyer, J.E. Effects of Metaphylaxis on Production Responses and Total Antimicrobial Use in High-Risk Beef Calves. *Appl. Anim. Sci.* **2020**, *36*, 265–270. [CrossRef]
21. Word, A.B.; Ellis, G.B.; Holland, B.P.; Streeter, M.N.; Hutcheson, J.P. Effects of Antimicrobial Metaphylaxis Using No Antimicrobial, Tilmicosin, or Tildipirosin and 2 Different Days on Feed on the Health and Growth Performance of Lightweight Beef Steer Calves Originating from Mexico. *Appl. Anim. Sci.* **2021**, *37*, 207–216. [CrossRef]
22. Rojas, H.A.; White, B.J.; Amrine, D.E.; Larson, R.L.; Capik, S.F. Associations of pen management characteristics and bovine respiratory disease incidence in the first 45 days post-arrival in feedlot cattle. *Bov. Pract.* **2022**, *56*, 40–52. [CrossRef]
23. Smith, K.; Amrine, D.E.; Larson, R.L.; Theurer, M.E.; White, B.J. Determining relevant risk factors associated with middle-day and late-day bovine respiratory disease morbidity in cohorts of cattle. *Appl. Anim. Sci.* **2022**, *38*, 373–379. [CrossRef]
24. Abell, K.M.; Theurer, M.E.; Larson, R.L.; White, B.J.; Apley, M. A Mixed Treatment Comparison Meta-Analysis of Metaphylaxis Treatments for Bovine Respiratory Disease in Beef Cattle. *J. Anim. Sci.* **2017**, *95*, 626. [CrossRef] [PubMed]
25. Horton, L.M.; Depenbusch, B.E.; Dewsbury, D.M.; McAtee, T.B.; Betts, N.B.; Renter, D.G. Comprehensive Outcomes Affected by Antimicrobial Metaphylaxis of Feedlot Calves at Medium-Risk for Bovine Respiratory Disease from a Randomized Controlled Trial. *Vet. Sci.* **2023**, *10*, 67. [CrossRef] [PubMed]
26. Dennis, E.J.; Schroeder, T.C.; Renter, D.G. Net return distributions when metaphylaxis is used to control bovine respiratory disease in high health-risk cattle. *Transl. Anim. Sci.* **2020**, *4*, 1091–1102. [CrossRef] [PubMed]
27. CattleFax-450 LB Steers. Available online: https://www.cattlefax.com/#!/data/cattle/cf-prices/450-lb-feeder-steer/ (accessed on 27 June 2023).
28. CattleFax-550 LB Steers. Available online: https://www.cattlefax.com/#!/data/cattle/cf-prices/550-lb-feeder-steer/ (accessed on 27 June 2023).
29. CattleFax-650 LB Steers. Available online: https://www.cattlefax.com/#!/data/cattle/cf-prices/650-lb-feeder-steer/ (accessed on 27 June 2023).
30. CattleFax-750 LB Steers. Available online: https://www.cattlefax.com/#!/data/cattle/cf-prices/750-lb-feeder-steer/ (accessed on 27 June 2023).
31. CattleFax-850 LB Steers. Available online: https://www.cattlefax.com/#!/data/cattle/cf-prices/850-lb-feeder-steer/ (accessed on 27 June 2023).
32. CattleFax Fed Steer Price. Available online: https://www.cattlefax.com/#!/data/cattle/cf-prices/fed-steer/ (accessed on 27 June 2023).
33. Valley Vet Supply. Available online: https://www.valleyvet.com/?ccd=ISM001 (accessed on 27 June 2023).
34. Waggoner, J. Focus on Feedlots Monthly Reports. Available online: https://www.asi.k-state.edu/about/newsletters/focus-on-feedlots/monthly-reports.html (accessed on 27 June 2023).
35. Johanns, A. Prices and Profitability Models. Available online: https://www.agmrc.org/renewable-energy/prices-and-profitability-models (accessed on 27 June 2023).
36. Woolums, A.R.; Loneragan, G.H.; Hawkins, L.L.; Williams, S.M. Baseline Management Practices and Animal Health Data Reported by US Feedlots Responding to a Survey Regarding Acute Interstitial Pneumonia. *Bov. Pract.* **2005**, *39*, 116–124. [CrossRef]

37. Zinn, R.A.; Barreras, A.; Owens, F.N.; Plascencia, A. Performance by Feedlot Steers and Heifers: Daily Gain, Mature Body Weight, Dry Matter Intake, and Dietary Energetics. *J. Anim. Sci.* **2008**, *86*, 2680–2689. [CrossRef] [PubMed]
38. Sanderson, M.W.; Dargatz, D.A.; Wagner, B.A. Risk Factors for Initial Respiratory Disease in United States' Feedlots Based on Producer-Collected Daily Morbidity Counts. *Can. Vet. J.* **2008**, *49*, 373–378. [PubMed] [PubMed Central]

animals

MDPI

Article

Application of Epidemiological Methods in a Large-Scale Cross-Sectional Study in 765 German Dairy Herds—Lessons Learned

Roswitha Merle [1,*], Martina Hoedemaker [2], Gabriela Knubben-Schweizer [3], Moritz Metzner [3], Kerstin-Elisabeth Müller [4] and Amely Campe [5]

1 Institute of Veterinary Epidemiology and Biostatistics, School of Veterinary Medicine, Freie Universität Berlin, Königsweg 67, 14163 Berlin, Germany
2 Clinic for Cattle, University of Veterinary Medicine Hannover, Foundation, Bischofsholer Damm 15, 30559 Hannover, Germany; martina.hoedemaker@tiho-hannover.de
3 Clinic for Ruminants, Ludwig-Maximilians-Universität Munich, Sonnenstr. 16, 85764 Oberschleißheim, Germany; g.knubben@lmu.de (G.K.-S.); m.metzner@lmu.de (M.M.)
4 Ruminant and Swine Clinic, School of Veterinary Medicine, Freie Universität Berlin, Königsweg 65, 14163 Berlin, Germany; kerstin-elisabeth.mueller@fu-berlin.de
5 Department of Biometry, Epidemiology and Information Processing (IBEI), WHO Collaborating Centre for Research and Training for Health at the Human-Animal-Environment Interface, University of Veterinary Medicine Hannover, Foundation, Buenteweg 2, 30559 Hannover, Germany; amely.campe@tiho-hannover.de
* Correspondence: roswitha.merle@fu-berlin.de

Simple Summary: The "PraeRi" study, conducted by three German veterinary universities from 2016 to 2020, was aimed at enhancing dairy health and welfare. With 765 visited dairy farms and 101,307 animals examined, the designed study provided an opportunity for improving animal well-being. Researchers applied good epidemiologic practice and concepts proven in large-scale studies, in a study comprising three different regions in Germany with varying farm characteristics. A sample size of 250 farms per region, determined through stratified sampling based on farm size, ensured representative estimates. According to the information provided by the farmers, mastitis was the most frequently occurring disease in their herds (14.2% to 16.3% of the herd—depending on the region). For most disorders, prevalence data were lowest for the region South compared with the two remaining regions. Statistical analyses identified risk factors, with results communicated through individual reports and benchmarking flyers to participating farmers. Challenges arose from managing practical procedures and communication due to the project's vast scale. Harmonizing data management and hypothesis testing across involved parties added complexity. Despite challenges, the PraeRi study considerably contributed to advancing dairy health and welfare practices.

Abstract: From 2016 to 2020, the "PraeRi" study, conducted by three German veterinary universities, was aimed at enhancing animal health and welfare in dairy farms. With 765 dairy farms visited and 101,307 animals examined, this study provided a basis for improving animal health and welfare. The study population comprised three different regions representing a broad variety of characteristics. To ensure representative estimates, a sample size of 250 farms was determined for each region, employing a stratified sampling plan based on farm size. According to the information provided by the farmers, the most commonly occurring disease in their herds was mastitis without general disorder (14.2% to 16.3% of the herd—depending on the region). For most disorders, prevalence data were lowest for the region South compared with the two remaining regions. Multivariable regression analyses were performed to identify risk factors for various target variables, and the results were communicated through individual reports and benchmarking flyers to participating farmers. The authors encountered challenges in management and communication due to the project's size in terms of personnel, data, and farms examined. Harmonizing data management and hypothesis testing across all involved parties added complexity.

Keywords: good epidemiologic practice; dairy cows; cross-sectional study; incidences of chronic diseases; dairy farms

1. Introduction

From 2016 to 2020, the large-scale study "PraeRi" (animal health, hygiene and biosecurity in German dairy herds—a prevalence study) was carried out by three German veterinary universities: University of Veterinary Medicine Hannover (North), Freie Universität Berlin (East), and Ludwig-Maximilians-Universität Munich (South). The overall goal of the study was to improve animal health and welfare in dairy farms. The main task was to provide reliable and representative estimates of basic dairy cow health indicators, such as mastitis, lameness, infectious diseases, etc., because prevalence estimates had not been available at that time. The latter data form an ideal base for strategies to improve animal health and welfare.

Three clinics and two epidemiologic institutes participated in the study. Eleven senior researchers and 43 study veterinarians worked on the project. In total, 765 dairy farms were visited, and data were collected (interviews with 376 questions; standardized measurements of stalls; examination of 186,160 animals; laboratory analyses) and recorded in 35 tables with 1522 variables in a relational SQL-database. The study was funded by the German Federal Ministry of Nutrition and Agriculture with a total amount of more than EUR 4.6 m (support codes: 2814HS006 (Hannover), 2814HS007 (Berlin), 2814HS008 (Munich)). The final report to the ministry consisted of a main document with 262 pages and eight appendices enclosing more than 276 pages (https://ibei.tiho-hannover.de/praeri/pages/69 (accessed on 31 March 2023)).

Researchers do not often have the chance to be part of a large-scale epidemiologic observational study because such studies are expensive and time-consuming. Only few funding bodies such as governmental research institutions are able and willing to invest large amounts of money into studies providing medium- and long-term results.

Large observational studies offer many possibilities to collect representative and reliable data on complex multifactorial issues. Dairy cow health disorders originate from complex interactions of housing conditions, herd management, and health management, as well as the farmers' personality and characteristics of the individual animal. The collection of data that include a high number of farms and cows as well as many variables is a precondition for sound and reliable statistical modeling.

Good epidemiologic practice (GEP) is a guideline for the state-of-the-art conduction and analysis of epidemiologic studies [1]. Large-scale studies have a higher chance of applying these guidelines than smaller ones.

Large-scale studies, however, pose challenges that do not occur in smaller studies. Data collection on a large number of farms including plenty of variables requires good study-management. As an example: when various observers collect data, good interobserver agreement must be safeguarded by repeated trainings. In addition, the analysis of the data is not as straightforward as in smaller studies with one or few target variables and less hierarchical structures. An analysis strategy must be agreed on by all researchers involved, and future use of the data must be coordinated.

We use our experiences from the study "PraeRi" to present and discuss an approach on how GEP and standard epidemiologic methods can be successfully applied to a large-scale study. Since this study was an observational study, we followed the guidelines of the STROBEvet statement [2] to structure the following text.

2. Materials and Methods

We conducted a cross-sectional study addressing animal health, hygiene, and biosecurity to achieve representative and reliable prevalence estimates of animal health disorders in German dairy farms.

2.1. Study Population

Regionalization: Three structurally different regions were studied, all of which make a significant contribution to milk production in Germany, but were different concerning farm size distribution and farm management. The latter differences were already reported by one of the co-authors in 2012 [3].

2.1.1. Sample Size of Farms per Region

Different target variables: We intended to calculate one (overall) sample size suitable for several different target variables. Hence, it was to be considered that each target variable had a different distribution (point estimate and standard deviation), when estimated on a farm level (i.e., distribution of milk yield per farm), or was measured as a percentage for the presence of a characteristic. As most of the target variables were originally collected on an animal level (i.e., quantitative value or dichotomous value for the presence or absence of a disease), these had to be aggregated on a farm level to obtain the statistical unit as the farm. For those target variables that were quantitative in nature, a measure of central tendency per farm was calculated. The distribution of this measure of central tendency was the target value to be interpreted in terms of content. For those target variables that were dichotomous in nature (e.g., presence of disease), the prevalence was calculated as the percentage of animals affected by the disease in question.

One for all sample size: To calculate the sample size, we took estimates from a previous study as a basis. First, we calculated sample sizes between 10 and 297 for quantitative target variables given a precision of 1 to 5 and a standard deviation of 7 at a confidence level of 0.95 to 0.99. Then, we calculated sample sizes between 19 and 2536 for dichotomous variables given a prevalence from 0.05 to 0.5 and a precision from 0.025 to 0.1. All calculations were conducted with NCSS PASS version 13.0.8 and under the consideration of Glaser and Kreienbrock [4]. Based on these scenarios and with feasibility in mind, a sample size of 250 farms was determined for each region.

2.1.2. Farm Size as Stratifier

To prevent a biased study population with regard to farm sizes, a sampling scheme was developed in which stratification was made according to small, medium, and large farms (control of selection bias). Farm sizes were chosen as cut-offs for this categorization, which allowed the target population to be assigned to three equally sized groups. These cut-offs were determined individually for each study region and transferred to the study population. Consequently, 83 farms were to be investigated per farm size category and region.

2.1.3. Sampling Procedure

A random sample of farms was drawn from the National Traceability and Information System for Animals "Herkunftssicherungs- und Informationssystem für Tiere" (HIT; https://www.hi-tier.de/ (accessed on 31 March 2023)) [5], with selection stratified by state or region and farm size. Based on the experiences of a previous study [6], a response rate of approximately 30 to 40% was anticipated.

For each region, 1250 farms were randomly selected, 5 times more than what was needed to cover a planned participation rate of at least 20%. Region North included the federal states of Schleswig-Holstein and Lower Saxony. Region East was represented by the states of Mecklenburg-Western Pomerania, Brandenburg, Saxony-Anhalt, and Thuringia, and Bavaria was studied for region South.

In Bavaria, the use of the HIT data was not approved by the Bavarian State Ministry for the Environment and Consumer Protection. Alternatively, the sampling could be performed using the database of the Milchprüfring Bayern e.V. (MPR; https://www.mpr-bayern.de/en (accessed on 31 March 2023)). Here, about 90% of the dairy farms located in Bavaria were registered.

Due to the low overall participation rate and the unintended inclusion of cattle farmers without dairy cows in the selection population, an in-depth analysis of the composition of the target population and the selection population from HIT was conducted in summer and early fall 2017. The objective was to (1) initiate a second draw of address data and, in doing so, (2) better reflect the true distribution of farm sizes of dairy farms. Therefore, the HIT animal data were also cross-checked with data from the federal state control associations (Dairy Herd Improvement associations (DHI), [7]). For this purpose, the September 2016 data were used to ensure comparability with the HIT data, which were also from September 2016.

Finally, a different type of address data management was defined for each region:

- Region North (N): The farm size classes were recalculated using the DHI information on farm sizes in Lower Saxony and Schleswig-Holstein and applied to the HIT data. The farm size information came from DHI data, while the address data drawing continued to be collected in HIT.
- Region East (E): Here, too, new farm size classes were determined using DHI data from Brandenburg, Saxony-Anhalt, Thuringia, and Mecklenburg-Western Pomerania. The address data were then also drawn from DHI data. A change in the address data source in the region East was justified by the fact that a very high proportion of dairy cow farmers (mean: 55%) and dairy cows (mean: 90%) were also members of a state control association in the states concerned. The advantage of writing exclusively to dairy cow farmers thus outweighed the disadvantage of not reaching those dairy cow farmers who were not members of the state control association. Due to the change in address data management in region East, it was unavoidable that farms were contacted twice with the second sampling. However, the DHI excluded those farms that had already participated in the study in advance.
- Region South (S): The calculation of the farm size classes was based on the data of the MPR. First and second data draws were made from the address data of the MPR, since access to address data from HIT had not been granted in Bavaria.

The finally realized farm size classes are shown in Table 1, and the final participation rate per region and class in Table 2 and Figure 1.

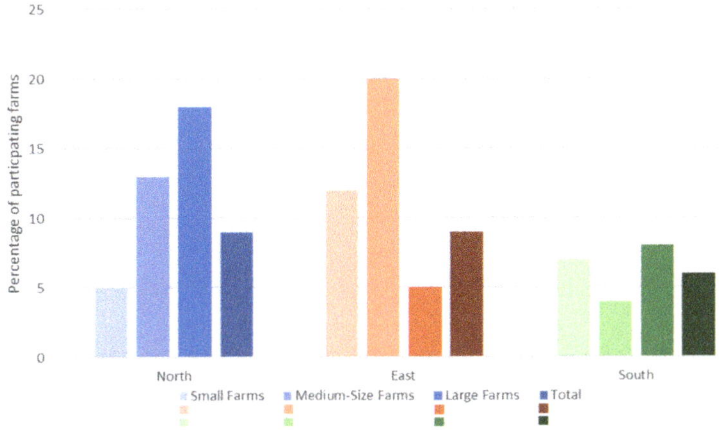

Figure 1. Overview of participation rate by region and farm size category in a prevalence study among 765 German dairy herds. Displayed is the percentage of farms that participated from all farms that were invited.

Table 1. Adapted cut-offs (number of cows) based on data originating from state control associations as opposed to data from the National Traceability and Information System for Animals in a prevalence study among 765 German dairy herds (data as of September 2016).

		Cut-Off	
Region	**Small**	**Medium**	**Large**
North	1–64	65–113	≥114
East	1–160	161–373	≥374
South	1–29	30–52	≥53

Table 2. Overview of participation rate by region and farm size category in a prevalence study among 765 German dairy herds.

	Number of Small Farms		Number of Medium-Sized Farms		Number of Large Farms		Total	
	Invited	**Visits**	**Invited**	**Visits**	**Invited**	**Visits**	**Invited**	**Visits**
Schleswig-Holstein	330	13	210	31	110	25	650	69
Lower Saxony	1334	70	464	59	339	55	2137	184
North	1664	83/5%	674	90/13%	449	80/18%	2787	253/9%
Mecklenburg-Western Pomerania	189	18	123	26	264	22	576	66
Saxony-Anhalt	156	20	131	26	82	26	369	72
Brandenburg	109	24	103	22	173	19	385	65
Thuringia	247	20	76	13	86	16	409	49
East	701	82/12%	433	87/20%	605	83/5%	1739	252/9%
Bavaria	1345	92	2015	84	1058	84	4418	260
South	1345	92/7%	2015	84/4%	1058	84/8%	4418	260/6%

2.1.4. Sampling of Animals per Farm

On large-scale farms of region East, it was not possible to examine all animals on a farm. The sample size for each farm was calculated to estimate an expected prevalence of 40% at a confidence level of 95%, with a power of 80% and a precision of ±5%. Thus, no more than 130 animals were sampled per farm in region South and no more than 213 in region North. In region East, all animals were sampled in farms with up to 159 animals, 166 in farms with 160 to 292 animals, and no more than 292 animals in farms with 293 animals or more [8]. The methods to ensure an unbiased selection of the cows evenly distributed over compartments were set out in the project's SOP manual.

For the calves, a sample calculation per farm was also performed. Here, an expected prevalence of 40% was to be estimated at a confidence level of 95%, with a power of 80% and an accuracy of ±10%. Consequently, no more than 33 calves were examined in region South and no more than 54 in region North. In region East, all animals were sampled up to a farm size of 40, 40 in farms with 41 to 73 animals, and from 74 calves onwards exactly 73 calves were sampled.

2.2. Questionnaires and Survey Forms

Based on the examination catalog from a previous project, as well as on literature research and expert opinions, the questionnaires and survey forms for the herd examinations were created and validated. The process of variable selection and question formulation was based on pre-formulated hypotheses, which were specifically linked to the disorders and performance features to be investigated. The questionnaires were used to record the relevant disease and performance characteristics of dairy cows, calves, and young stock in Germany, as well as potential factors influencing them (Supplemental Material S1). The survey forms were used to investigate cows (e.g., body condition score [9], locomotion score [10,11], and hygiene score [12]). A new tool for the assessment of hygiene in calves was developed and published by Kellermann et al. [13].

It was determined by which method, from which source, at which level, and for which animal groups information should be collected. Feasibility, validity, and repeatability were considered. An internal and external validation of the questionnaires and survey forms was performed via testing in pilot farms (South: n_{farms} = 5, $n_{animals}$ = 357; East: n_{farms} = 3, $n_{animals}$ = 944; North: n_{farms} = 6, $n_{animals}$ = 1424). Based on the experiences from the pilot phase, the questionnaires and survey forms were modified once again. In the end, 15 questionnaires and 15 survey forms were created, some of which had to be filled out only once, and some of which had to be filled out several times (e.g., for all different silages in use).

2.3. Database and Homepage

A homepage (www.PraeRi.de (accessed on 31 March 2023)) with information on the project, its objectives, the study teams, and the status was available for animal owners and veterinarians. It is still operated and can be used to look up results (including the final report).

An SQL database with a web interface was set up with the scripting language PHP and maintained on a virtual server at the University of Veterinary Medicine Hannover Foundation, Germany. There were different user groups, and the data input could be conducted and decentralized. All data of the project could be viewed by all users according to their user rights. Functions were available for general, as well as individualized, export of data. After the creation of the database and input interfaces, an internal (with colleagues from the IBEI) and external (with colleagues from study team North) check was carried out and any errors were corrected, or missing items added. The database comprised 35 main tables, 132 check tables, and 1522 data fields. It was used by about 40 users with different data access and user rights. In addition to the data from the questionnaires and survey forms, most of which had to be entered manually by the study veterinarians, import functions were available, which were also set up for data from external sources (e.g., HIT (www.hi-tier.de (accessed on 31 March 2023)), DHI, and LUFA Nord-West (accredited service laboratory of the Lower Saxony Chamber of Agriculture, https://www.lufa-nord-west.com/ (accessed on 31 March 2023)) [14], ration key figures, parasitology laboratory). Throughout the reporting period, various requests were raised by the study teams, errors were uncovered, and changes and additions were made. Therefore, various additional programming was carried out to facilitate the use of the database, data entry and data correction. Furthermore, explanations on the use of various database parts were available (user manual).

2.4. Farm Recruitment

Employees of the HIT database created a sample according to the above-mentioned sampling requirements. The sample consisted of an address list and was sent to the top state representatives of the veterinary authorities to preserve data protection. They passed it on to the study units responsible for address data management. In the region East, this was executed by the regional state control associations for milk testing. The latter institutions ensured the address data were administered and established contacts with the farmers asking if they were willing to participate. In Schleswig-Holstein, this was carried out by the Ministry for Energy Transition, Agriculture, Environment and Rural Areas, and in Lower Saxony by the project secretary of the Clinic for Cattle at the University of Veterinary Medicine Hannover. Since the Bavarian State Ministry for the Environment and Consumer Protection did not agree to the use of HIT data for farm recruitment in Bavaria due to data protection reasons, the random sampling and printing of the cover letters (see below) in region South was carried out at the MPR.

Uniform cover letters were prepared for the participants in the three regions. The selected farms received such a cover letter by postal mail together with study-specific information (in the form of a flyer), a reply postcard, and a postage-paid envelope. Persons who wished to participate in the study returned the reply postcard to the respective regional office (by postal mail, fax, or e-mail).

When receiving the farmers' positive reply, the study veterinarians gained access to the contact data for the first time. They made an appointment for a telephone interview, in which the contents of the study and the investigations planned on the farm were explained to the farm managers. Farms with farm-gate sales and farms which delivered their produced milk to neighboring countries were not included. Then, they requested information required to prepare adequately for the farm visit. At the end of the telephone interview, an appointment was made for the visit. Farmers were also informed that they must ensure access to their HIT- and DHI-databases during the farm visit, so that the previous year's data could be downloaded.

2.5. Quality Assurance Measures

Several measures were taken to ensure that data were collected consistently across all three study regions, to the extent possible.

2.5.1. Sampling of Animals per Farm

Through the intensive exchange during the development of the questionnaires and survey forms, all study veterinarians benefited from the different experiences that the individuals contributed to the project. At the same time, the literature research and the exchange with additional experts led to a familiarization with the different topics.

2.5.2. Training to Enhance Interobserver Reliability (IOR)

For quality assurance, the study veterinarians were introduced and trained in livestock surveys (three-day training in September 2016).

After the completion of the pilot phase of the project, a one-day alignment of different measurements took place in Hannover at the university's Teaching and Research Farm in 2016. The aim of the comparison was to determine whether and to what extent the study veterinarians (observers) agreed in their assessment of the animals. This was conducted to minimize the possible effects of individual observers and thus to be able to collect the PraeRi data in a harmonized way. The criteria for locomotion and hygiene assessment, as well as the measurement of cow sizes were fulfilled. To identify substantial deviations in single observers, an exclusion test was performed according to Ruddat et al. [15]. In total, there was already good to very good agreement for some measurements at the beginning of the study. For recalibration of the study veterinarians, two further IOR comparisons took place. Systematic regional effects could not be identified in this observer assessment.

At the second training in 2017, 14 calves and 60 cows were evaluated by each of the 16 study veterinarians present. The third training was held in Munich in 2018. During this two-day training, 60 cows and 20 calves were evaluated again by the 15 study veterinarians present. Additionally, 13 cows in tethered housing were assessed. Furthermore, photos or videos were evaluated and discussed jointly by all study veterinarians in the context of a group discussion for distinct measurements, which had emerged as problematic in the previous training sessions.

The data of the assessment of the cows and calves were analyzed, and the results were communicated in a general report to all project partners. In addition, observers whose results significantly deviated from the majority of study veterinarians with respect to at least one observation received a personal notification which enabled them to adjust their assessment in the future. An analysis of the data from the second and third comparison showed that there was an overall satisfactory agreement between the study veterinarians.

Except for differences in the auscultation of the calves, no evidence of region-specific differences could be found. Due to partly only very small differences between the examined animals, the agreement between the study veterinarians could not be evaluated with the desired certainty for all characteristics.

2.5.3. SOP Manual

To ensure a uniform procedure for the collection of data by the study veterinarians, standard operating procedures (SOPs) were developed and summarized in a collection of methods (SOP manual). In addition to the SOPs for the individual questionnaires, the manual also contained a glossary to explain the special terms used to the farmers and standard procedures for filling out the questionnaires. Any ambiguities that arose during the study regarding the collection of data were discussed in regular telephone conferences of the study veterinarians and corresponding decisions were documented in the manual.

2.5.4. Leaders Video Conferences and Collaborative Meetings

Video conferences of the leaders: Video conferences of the project managers of the three study teams took place in monthly intervals. These were used to exchange information on the current status of the studies, to discuss and solve problems that had arisen, and to agree on the next steps in the project.

Consortium meetings and coordination of the analyses: a consortium meeting was held in Berlin in 2018 to coordinate the work in the context of data analysis, of the preparation of the final report, and the scientific publications of the overall results. Eight thematic working groups were formed that prepared detailed descriptive evaluations and the appropriate statistical models (see below for more information on the approach). Furthermore, a two-day collaborative meeting for the actual analyses took place in Hannover in 2019, where the two epidemiological institutions presented detailed instructions for the steps of the analyses.

Telephone conferences of the study veterinarians: In addition, telephone conferences between the study veterinarians and the epidemiological institutes were held at approximately bi-weekly intervals. In these, mainly emerging issues or peculiarities observed on the farms were discussed and a decision was made on how to handle such cases. Questions that remained unresolved were discussed and, if necessary, final decisions were made in the monthly leaders' video conference. All decisions made during telephone and video conferences were subsequently included in the manual.

2.6. *Data Analysis*

Data were pseudonymized, i.e., the identification patterns such as the name and address of the farm were stored separately. The data set used for analyses did not contain any data that would enable third-party member to identify the farms. The list for re-pseudonymization was only accessible to the regional study teams in order to send feedback to the farmers.

2.6.1. Feedback Letters to Farmers

In total, two feedbacks were sent to the participating farmers. The first letter represented an individual report that was prepared a short time after the farm visit and included an overview of all results of the investigations: It included the evaluations of the skin lesions as well as of the hygienic conditions, a number of tables and figures displaying the occurrence of lameness, the distribution of the body condition score, and an overview of the calf weight related to age. An example of this letter is given in Supplemental Material S2.

After total completion of the data collection process and baseline descriptive analyses, another round of feedback was sent to all participating farmers in the form of a benchmarking flyer, in which individual farm data were compared with the distribution of data of all study participants (see Supplemental Material S3).

2.6.2. Plausibility Controls

The epidemiology groups performed general plausibility checks on approximately 230 quantitative variables from the interview questionnaires and approximately 50 variables from the data entry forms. For these variables, no database internal validations during data entry were applied, which would have caused error messages in the case of non-plausible entries. Therefore, the observers went through a number of lists with potentially implausible values and checked questionnaires and survey forms for the correct values. Accordingly, approximately 130 related questions (so-called filter questions) were checked for logical relationships. When these respective data were entered into the database, a variable-specific, logic-based pre-assignment of dependent answers initially took place automatically. However, this could be changed by the data entry person if needed, which made it necessary to check the data records after data entry. Following the plausibility checks, database queries were defined and programmed to allow the study teams to retrieve clean and summarized data sets.

In addition, database queries were created that related to specific questions. On the one hand, these were adapted from the planned analyses in the working groups. On the other hand, imported data from the external data sources HIT and DHI had to be checked very intensively. Some information was included in both data sets. The information, however, included was not always consistent. For example, information on the lactation number of animals did not match the recorded number of calvings. The reasons for such discrepancies (here, for example, the non-reporting of stillbirths in HIT) were clarified. We identified which data source had the greatest completeness and reliability for the respective information. With the help of contact persons at HIT and "Vereinigte Informationssysteme Tierhaltung w.V." (https://www.vit.de/en/ (accessed on 31 March 2023)), discrepancies could be traced, and further work could be conducted with these findings in the best possible way.

All plausibility checks required intensive exchange between the study teams. Expertise and experience from previous farm visits were considered, as well as region-specific features. For the specific plausibility checks, the epidemiologists sent lists of data sets to the study teams, which then checked these and made necessary corrections in the database. All plausibility checks resulted in a very low number of missing values, on the one hand, and an extremely high quality of the available data, on the other hand.

2.6.3. Multivariable Regression Analyses

Working groups: As described above, topic-specific working groups were formed to guide the inductive statistical analyses. Each group consisted of study veterinarians from all three regions, an epidemiologist and a project leader.

The following seven topics were addressed in this way:

- Udder Health;
- Reproduction;
- Metabolism;
- Limb health and lameness;
- Technopathies such as skin lesions;
- Calves/young cattle;
- Feeding;
- Infectious diseases/biosecurity.

Communication within the working groups took place via telephone (bilateral/conference), video conference, or e-mail. In addition, internal working group meetings were held at one of the study sites on an irregular basis. The task of the individual working groups was to define target values in their area and to assign potential risk factors. Hypotheses: A hypothesis-based approach was chosen to select the variables for each model. First, hypotheses were formulated including one target variable (outcome) and one or several influencing factors. The whole consortium agreed on a summarized list of hypotheses.

In the next step, causal diagrams were created with the help of a hypothesis list. After identifying the target variable, a causal directed acyclic graph (DAG) (http://www.dagitty.net/ (accessed on 25.05.2023)) [16] was created, including all known and expected influence factors on this outcome, from existing and/or expert knowledge. Drawing arrows between the variables enabled the scientists to identify confounders (common causes for influence factor and outcome) as well as variables causing selection or measurement bias when analyzing the effect of one specific variable on the outcome. An example of a causal diagram is shown below (Figure 2). For the statistical analyses, some of the influence factors were latent and were, therefore, represented by observed variables in the data collected.

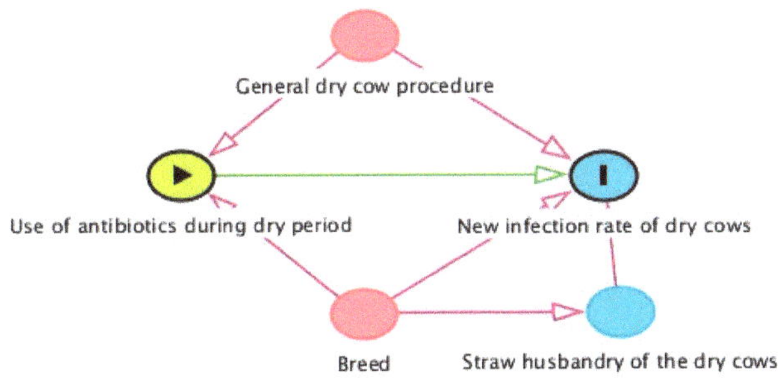

Figure 2. A causal diagram displayed as a directed acyclic graph in a prevalence study among 765 German dairy herds showing the example target variable "New infection rate of dry cows", its influencing variable "Use of antibiotics during the dry period", and the possible confounders.

Example Clinical udder Infections

Hypothesis: Cubicle cleanliness has an influence on the incidence of clinical udder infections in such a way that dirty cubicles negatively affect udder health.

Target: Frequency of clinical udder infections, e.g., in the last 12 months before the farm visit related to the number of dairy cows (lactating and dry cows) in this period.

Influence variable: "Cleanliness of cubicles":

- Step 1: Define the animal group under consideration, e.g., lactating.
- Step 2: Definition of the hierarchical level, e.g., at the plant level, at least one compartment with the grade "soiled" [yes/no].
- Step 3: Description of influencing variable, e.g., farm has at least one dirty compartment with lactating dairy cows yes/no.

In most cases, the statistical unit was the farm, but the data were sometimes also collected at the animal, compartment, barn, and possibly site level. Data agglomeration at the various levels was therefore necessary, i.e., "percentage of cows with specific characteristics" (e.g., lameness) or "predominant category present in at least 80% of the cows" (e.g., breed). It became clear that the preparation of prudent sets of variables needed the expertise of both the study veterinarians and the epidemiologists.

2.6.4. Statistical Analyses

Data from different levels of organization from one farm, such as an animal, stable, and farm, were linked via the farm ID.

Descriptive statistics: For qualitative variables, both the absolute number and the percentage of study units (farms, compartments, animals...) per expression (category and response option) were presented. This was conducted separately by the study region. Quantitative variables were considered based on their distributions (measures of central tendency and variation). Variables at lower hierarchical levels (e.g., barn, compartment, or animal level) were often analyzed at their level of origin as well as at the farm level. For the presentation of qualitative variables at the farm level, the percentage of each characteristic per farm was calculated. The distribution of these percentages was summarized per region. For quantitative variables, this distribution was calculated for an arithmetic mean and median.

The mean percentage of cows leaving the farm was defined as the percentage of lost animals out of all animals on the farm in the year before the farm visit. The culling rate considered the days an animal was present during the study period and provided additional information about the "rate" at which an animal was likely to leave the farm as a result of sale for breeding or slaughter. For example, a mean culling rate of 52.2 per 100 cows per year means that the probability of a cow leaving after a full year is 52.2%, or, in other words, that on average each cow will leave within about 700 days. The loss rate was calculated as

$$\text{number per 100 cows and year} = (\text{number of lost animals}/\text{number of animal days in the period}) * 365 * 100 \quad (1)$$

Information on reasons for leaving was obtained from the DHI and HIT data. From the DHI data, voluntary and involuntary reasons for cows leaving the farm were derived. When evaluating the figures, it must be considered that in a substantial number of cases, not single but multiple reasons led to the decision that the animal had to leave the farm.

The mortality rate was calculated as

$$\text{number per 100 cows per year} = \text{Number of animals died/killed}/\text{number of animal days in the period}) * 365 * 100 \quad (2)$$

Regression models: Generalized regression models were performed for each target variable to identify relevant risk factors for the target variables of interest (e.g., disease frequencies and skin lesions). If the target was on any other than farm level, a hierarchical model including the farm, and, if necessary, barn level as random factors was developed.

During data management, all variables required for the respective model were compiled. The documentation of the data management was conducted in an Excel file containing information on variable labels, data type, coding, and clear names of the variables, as well as a reference to the question in the questionnaire. The members of the working groups jointly determined which variables would be transformed (e.g., take the logarithm if there was no normal distribution) or recategorized (combine categories). Missing data were labeled with special missing values such as −77 (not applicable), −88 (do not know), and −99 (not specified).

First, a univariable generalized regression model was created separately for each influencing factor (univariable evaluations). In addition, a bivariate correlation analysis of all variables among each other was performed. The result of these steps resulted in the selection of the influencing variables which were potentially included in the multivariable model:

- Potential association with target variable, expressed as
    - Correlation coefficient > 0.1; or
    - p-value < 0.2.
- Factors indispensable to the content.

If two influencing variables were highly correlated, the one with the lowest p-value was selected or the two variables were combined into one.

Multivariable Analyses

The confounder analysis was performed separately for each of the influencing factors. A causal diagram was now created using these variables. In our example, the new infection rate of dry cows is the target variable, and the use of antibiotics during the dry period is the influencing variable (Figure 2). General dry cow procedures and breed can influence both the target and influence variable. Therefore, these were considered confounders and had to be included in the model, if the actual influence on the target variable was to be investigated. Straw bedding of the dry cows, on the other hand, only influenced the target variable.

Following the respective DAG, the influencing factors including all confounding variables and those that represented sources of selection and measurement bias were included in one (hierarchical) generalized regression model which served as the final model for the relationship between one influencing variable and the outcome. This resulted in several models for one outcome.

Since this procedure is very time-consuming, we had to use a more pragmatic approach for some outcomes: First, a maximum model was created in which all influencing variables were included with their associated confounders. Then, the variables were selected stepwise backwards by removing the variable with the largest p-value or with the smallest loglikelihood (-2ll). This was continued until only variables with $p < 0.2$ were left in the model. However, the variables that were indispensable always remained in the model. In the next step, the respective confounders for these variables were reinserted. Finally, possible interactions of influencing factors were determined and tested. Only interactions with $p < 0.05$ remained in the model. The final model was, therefore, composed of

- indispensable factors in terms of content;
- factors with $p < 0.2$;
- their confounders from the causal diagram;
- interactions between two factors with $p < 0.05$.

Analyses were performed with SAS® Software, version 9.4 of the SAS system for Microsoft (SAS Institute Inc. 2019, Cary, NC, USA). We used Proc Tabulate, Proc Means, Proc Univariate, and Proc Box-plot for descriptive analyses, as well as Proc Reg, Proc GLM, and Proc Glimmix for the regression models.

3. Results

3.1. Study Population

Participation rate: Overall, a participation rate of 6–9% was achieved (with slight regional differences) (Figure 1).

Representativeness of the study population: In this study, data were collected from 765 farms, and 86,304 individual cows, 15,003 individual calves, and 84,853 young animals kept in groups were examined (Table 3). The greatest number of animals was examined in the region East, which was due to the fact that dairy farming in the latter region—in contrast to regions North and South—is characterized by larger dairy operations with bigger herd sizes.. In addition, data from HIT and the milk performance tests were downloaded and evaluated. Table 4 presents the numbers of farms and cows in Germany compared to the figures in the PraeRi study. Due to the different numbers of farms and cows per region, the percentage of farms and animals included in the study was considerably higher (11 and 15%, resp.) in region East than in North (2%) and South (1%).

Table 3. Number of animals examined in a prevalence study including 765 German dairy herds.

	Number of Examined Calves		Number of Examined Cows		Number of Examined Young Stock	
	Average/Farm	**Total**	**Average/Farm**	**Total**	**Average/Farm**	**Total**
North	15	3741	99	24,980	77	19,571
East	37	9188	198	49,936	222	56,058
South	8	2074	44	11,388	35	9224
Total	20	15,003	113	86,304	111	84,853

Table 4. Representativeness of the PraeRi study population compared with data from the stat. Bundesamt (Destatis/GENESIS database, status as of 1.3.2016) (PraeRi: prevalence study among 765 German dairy herds).

	Number of Farms		Number of Cows		% of Farms in Study	% of Cows in Study	Average Farm Size		Distribution of Farms in Region (%)		Distribution of Cows in Region (%)	
	Genesis	**PraeRi**	**Genesis**	**PraeRi**	**PraeRi**	**PraeRi**	**Genesis**	**PraeRi**	**Genesis**	**PraeRi**	**Genesis**	**PraeRi**
Lower Saxony	10,080	184	864,750	18,188	2	2	86	99	71	73	69	70
Schleswig-Holstein	4180	69	396,358	7829	2	2	95	113	29	27	31	30
North	14,260	253	1,261,108	26,017	2	2	88	103				
Brandenburg	539	66	159,964	24,884	12	16	297	377	24	26	28	29
Mecklenburg-Western Pomerania	712	72	180,918	25,504	10	14	254	354	32	29	31	29
Saxony-Anhalt	520	65	123,405	22,352	13	18	237	344	23	26	21	26
Thuringia	485	49	110,502	14,098	10	13	228	288	21	19	19	16
East	2256	252	574,789	86,838	11	15	255	345				
Bavaria	32,564	260	1,208,640	11,539	1	1	37	44	100	100	100	100
South	32,564	260	1,208,640	11,539	1	1	37	44				

3.2. Baseline Results

3.2.1. Study Population

A total of 765 farms (North: n = 253; East: n = 252; South: n = 260), roughly evenly distributed among the three farm size categories, were visited. The farm visits took place from early December 2016 to late July 2019. As outlined in Section 2, cows were evaluated using various grading schemes, and additional clinical examinations were performed on calves.

With respect to farm size, the mean number of lactating cows per farm was (minimum-maximum) 120 (North: 15–1165, n = 242), 396 (East: 3–3365, n = 249), and 51 (South: 6–254, n = 232), respectively. The mean number of dry cows was 17 (North: 1–225, n = 241), 67 (East: 1–492, n = 245), and 8 (South: 1–47, n = 228), respectively. It should be noted that only farms that participated in the DHI testing were considered here.

In all three study regions, the cows in the first lactation accounted for the largest proportion at approximately 30 to 35% per farm, followed by lactation numbers in descending order. Cows had a median lactation number of 2.6 (North), 2.5 (East), and 2.8 (South) lactations, respectively. Compared to the regions North and East, it is noticeable that cows in the region South potentially reached more lactations.

Of the cattle breeds listed in Appendix 6 of the Livestock Trade Regulations (2016), 23 were used as dairy cows on the farms visited, although some were beef cattle breeds. In the regions North and East, the German Holstein breed was the most common, with an average of 82.9% (Median: 95.0, SD: 28.0) and 84.1% (Median: 93.8, SD: 24.7) per farm, respectively, while in the region South, the Simmental breed was the most common, with an average of 80.4% per farm (Median: 100.0, SD: 35.1). In the region North, the order of other cattle breeds was Red Holsteins (Median: 2.1), cross-bred dairy cattle x dairy cattle (XMM), crossbred beef cattle x dairy cattle (XFM), and dual-purpose Red Holsteins. In the region East, the order of the other cattle breeds was XMM (Median: 3.9), Red Holsteins, other breeds, and Simmentals with also minor frequencies. In the South, the second most common cattle breed was Brown Swiss (Mean: 11.8%, Median: 0.0) followed by Holstein-Friesian (Mean: 3.9, Median: 0.0), other breeds, and XFM.

The proportion of farms participating in the DHI testing was very high at 90% and above, depending on the region. Compared to the region North (4.3%, n = 11) and the region East (1.2%, n = 3), the proportion of farms that did not participate in DHI testing was highest in the region South with 10.8% (n = 28). The mean annual milk yield (minimum–maximum) based on the DHI farm results from the last available audit year varied between 7606 kg (3712–10,598 kg, n = 231) in the South, 9055 kg (3597–11,927 kg, n = 241) in the region North, and 9250 kg (2739–12,907 kg, n = 249) in the region East, respectively.

3.2.2. Farm Structure

Almost all farms in the regions North and East were run on a full-time basis (North: 98.9%, n = 250; East: 97.2%, n = 245). In contrast, 20.4% (n = 53) in the region South were part-time farmers. More than 80% of the farms were conventionally farmed. There were more organic farms in the South than in the other two regions (North: 4.4%, n = 11; East: 9.1%, n = 23; South: 13.9%, n = 36). In the region South, six farms (2.3%) reported being in the process of converting from conventional to organic. In more than 80% of the farms, cows were milked with conventional milking systems. A milking robot was present in 19.9% of farms in the region North (n = 50), in 14.7% of farms in the region East (n = 37), and in 11.9% of farms in the region South (n = 31).

During the farm visit, different survey sheets were completed for loose housing and tethering, depending on the housing system. Initially, it was not considered whether a certain type of housing was the predominant type of housing. However, based on the number of cows evaluated, it was possible to find out which was the predominant housing system, i.e., in which more than 80% of the cows graded were kept on the day of the farm visit. Loose housing was the predominant housing system in regions North and East (North: 92.9% of farms, n = 235; East: 96.0%, n = 242). However, this was the case in only 61.2% of the farms in the region South (n = 159). Overall, the tethering of cows was considerably less common than loose housing, but was more frequent in the South compared with the regions North and East (North: 3.6% of farms, n = 9; East: 1.2%, n = 3; South: 29.2%, n = 77).

The farmed area was significantly larger in the region East (1059.1 ha on average) than in the North (106.2 ha) and South (49.8 ha) regions. While about half of the farmland in the regions North and South consisted of grassland, this was true for 20% of the farms in the region East.

In addition to milk production, most farmers reared their own youngstock (Table 5). In the regions North and East, approximately one-third of the farms kept mating bulls and/or fattening bulls. Another more frequent cattle-related farm activity in the region East was suckler cow husbandry. In contrast, calves and young cattle from other farms were raised rarely in general, however, more frequently in the region North than in the other two regions.

Table 5. Agricultural business branches besides dairying in a prevalence study among 765 German dairy herds.

Other Cattle Holdings	Region					
	North		East		South	
	n	%	n	%	n	%
Rearing calves	242	95.7	249	98.8	255	98.1
Breeding young cattle	239	94.8	237	94.0	243	91.9
Stud bulls	90	35.6	94	37.3	26	10.0
Bull fattening	79	31.2	57	22.6	27	10.5
Heifer fattening	23	9.1	19	7.5	13	5.0
Calf fattening	6	2.4	20	7.9	12	4.6
Suckler cow husbandry	9	3.6	51	20.2	3	1.2
Rearing calves from other farms of origin	29	11.5	13	5.2	10	3.8
Rearing young cattle from other farms of origin	26	10.3	16	6.3	10	3.8
No answer selected/does not apply	11	4.4	4	1.6	0	0.0
Total number of farms	253		252		260	

Summer grazing varied in different animal groups and depending on the region and farm size. Comparing regions, animals of all age/lactation stages in the region North were on pasture in summer on many farms (North: approx. 60% of farms, East: approx. 20%; South: approx. 30%). With increasing farm size, the offer for grazing decreased. In all three regions, grazing very often was provided to dry cows (North: 71.9%, n = 182; East: 56.0%, n = 141; South: 35.8%, n = 93).

3.2.3. Information on the Interview Partners

In the regions North and South, interview partners were owners and managers (North: 70.8%, n = 179; South: 81.2%, n = 211), predominantly. Equal partners were the second most interviewed group (North: 21.0%, n = 53; South: 11.9%, n = 31). In the region East, owners and managers (48.8%, n = 123) as well as herd managers and employees (42.5%, n = 107) were interviewed in equal numbers.

While the most common levels of education of the interviewees in the North were training as a master farmer (60.1%, n = 152) and training as a farmer (19.8%, n = 50), these training statuses were present equally often in the region South (master: 36.7%, n = 94; farmer training: 37.7%, n = 98). In contrast, the most common level of education in the region East was agricultural studies (59.9%, n = 150), followed by master farmer (17.1%, n = 43) and training as a farmer (14.3%, n = 36). It should be noted that, frequently, several levels of education were completed. The numbers given above, however, represent the highest level of education in case several levels were achieved. If we look at the level of training as a function of farm size, we notice a correlation in the regions North and East with higher levels of agricultural training, especially with a university degree. In the region South, where only a few interviewees had a university degree, a dependency on the size of the farm, although not as clear as in the other two regions, was only the case for training as a farmer and the acquisition of the title of master farmer.

3.2.4. Animal Health and Animal Health Management

Integrated veterinary herd management (IVHM) was applied by 54.1% (n = 137) of farms in the region North, 59.9% (n = 151) of farms in the East, and 18.1% (n = 47) of farms in the South. On average, IVHM was carried out in two to three different areas per farm. In all three regions, IVHM was most frequently implemented in the areas fertility (North: 84.7% of the farms applying IVHM, n = 116; East: 92.7%, n = 140; South: 85.1%, n = 40), and udder health (North: 51.1%, n = 70; East: 83.4%, n = 126; South: 42.6%, n = 20). In the North, control of lameness, young stock health, and nutrition were further areas in which IVHM was applied by one-third of the IVHM farms, respectively. In the region East, lameness control (55.0%, n = 83) and youngstock health (71.5%, n = 108) were IVHM areas in even more farms, while, especially, lameness control (6.4%, n = 3) was not a common IVHM area in the South as in the other regions.

The mean incidence of diseases is shown in Table 6. It must be noticed that these incidences were based on information provided by the animal owners. Some of the figures came from very precise farm documentation, but in many cases only from an estimate by the animal owners (approx. 50–70% of the animal owners depending on the disease). All prevalence distributions had a right-skewed distribution. According to the farmers' specifications, the most common disease in the herds was mastitis without general disorder (14.2% to 16.3% of the herd—depending on the region). For most diseases, prevalence data were lowest for the region South compared with the other two regions. Interestingly, in regard to foreign-body disease, mastitis with general disorder, and abortion, prevalence estimations did not differ considerably between regions.

Table 6. Mean incidence (%) of selected diseases in cows (based on farmers' interview data) in a prevalence study among 765 German dairy herds.

Variable	Region	Number of Farms	Mean	Standard Deviation	Me-Dian	25%-Quantile	75%-Quantile	Missing Values
Milk fever	North	250	10.6	10.0	7.4	4.3	14.5	2
	East	240	6.7	7.2	5.0	2.3	9.5	12
	South	257	5.8	5.7	4.7	1.7	8.3	3
Retained placenta	North	251	11.4	8.8	9.7	5.8	14.8	1
	East	236	10.2	8.1	9.2	5.0	13.3	5
	South	258	8.2	5.9	7.3	4.5	11.0	1
Uterine inflammation	North	252	8.8	10.3	5.4	2.6	11.1	1
	East	235	12.0	12.9	8.2	2.9	16.3	17
	South	257	4.5	6.0	3.3	0.0	6.7	3
Pneumonia	North	252	1.5	3.1	0.0	0.0	1.7	1
	East	239	2.3	6.1	0.8	0.0	2.0	13
	South	260	0.7	2.0	0.0	0.0	0.0	0
Ketosis	North	247	7.1	8.1	5.0	2.0	10.0	6
	East	234	6.3	8.3	3.0	1.0	9.0	18
	South	254	2.8	4.8	0.0	0.0	3.9	6
Displaced abomasum	North	253	2.1	2.2	1.6	0.0	3.0	0
	East	243	1.8	2.3	1.0	0.3	2.5	9
	South	260	0.3	1.1	0.0	0.0	0.0	0
Foreign body disease	North	251	1.4	2.5	0.0	0.0	2.0	2
	East	232	1.2	3.4	0.0	0.0	1.0	20
	South	259	0.9	3.2	0.0	0.0	0.0	1

Table 6. *Cont.*

Variable	Region	Number of Farms	Mean	Standard Deviation	Me-Dian	25%-Quantile	75%-Quantile	Missing Values
Mastitis without general disorder	North	252	16.3	11.3	14.8	8.7	21.5	1
	East	232	22.0	19.9	15.9	5.9	32.4	20
	South	259	14.2	12.5	11.8	6.0	19.2	1
Mastitis with general disorder	North	252	5.0	5.1	4.0	1.4	6.5	1
	East	228	4.9	8.0	2.0	1.0	5.0	24
	South	260	4.6	7.3	2.6	0.0	6.6	0
Heifer mastitis	North	253	3.4	4.6	2.1	0.0	4.4	0
	East	219	6.3	8.3	3.4	0.9	8.3	33
	South	225	2.5	5.6	0.0	0.0	3.3	5
Abortion	North	250	2.8	2.5	2.2	1.4	4.0	3
	East	239	2.9	2.9	2.0	1.0	4.0	13
	South	260	2.9	3.3	2.5	0.0	4.1	0
Lameness	North	252	12.6	12.1	9.5	4.8	16.4	1
	East	250	13.8	14.9	9.5	3.8	20.0	2
	South	259	8.5	7.8	7.1	3.3	11.5	1

Figure 3 shows that infertility, udder diseases, and disorders of the locomotor system were the main reasons for culling. While in the region North, the proportion of animals sold for breeding was significantly higher than in the other two regions, the South showed a high proportion of animals that left due to old age. In the East, on the other hand, a higher proportion of animals left due to underperformance than in the other two regions. In all regions, the high proportion of cows for which "miscellaneous" was recorded as the reason for culling was striking.

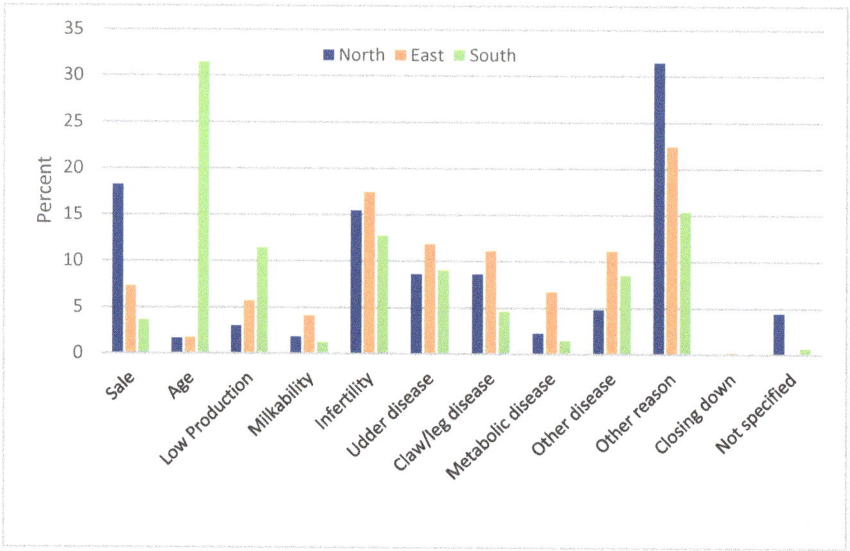

Figure 3. Reasons for cows leaving the farm due to sale or culling in a prevalence study among 765 German dairy herds. Shown is the mean proportion of cows (in% of cows that left) that left the farm for the reasons listed, staggered by region: North (n = 242 farms), East (n = 249 farms), and South (n = 232) farms (database: DHI).

The culling risk was similar in all three regions (North: Mean: 33.5, Median: 32.9, SD: 6.9; East: Mean: 37.5, Median: 35.9, SD: 9.5; South: Mean: 37.7, Median: 36.3, SD: 9.4). The mean culling rate in the region North (Mean: 52.2, Median: 49.5, SD: 18.2) was slightly lower than in the East (Mean: 64.9, Median: 55.6, SD: 32.9) and in region South (Mean: 65.7, Median: 57.3, SD: 31.3).

The mean percentage for mortality, i.e., the percentage of dead/killed animals out of all animals on the farm in the year prior to the farm visit, was 3.7% (North: median: 3.2, SD: 2.5), 4.2% (East: median: 3.9, SD: 2.2), and 2.3% (South: median: 1.8, SD: 2.1). In the region East, the median mortality rate was 6.3, 4.7 in the North, and, in the South, it was 3.4.

4. Discussion

To the authors' knowledge, a large-scale study, as reported in the present paper, has not been conducted on dairy cow health before. While the management of practical procedures and communication were challenging due to the sheer size of the project in terms of the amount of personnel, data collected, and farms and animals examined, we had to organize harmonized approaches for data management and hypothesis testing for all involved parties.

4.1. Sample Size

To receive valid estimates, the sample must be large enough and representative [17]. To find an optimal number of farms to be sampled, we started with calculating the sample size for several different target variables. This led to a multitude of different possible sample sizes, which was not satisfying. Thus, we needed to find a compromise between statistical necessity and feasibility in the field given the available time and personnel. From an epidemiological point of view, it must be concluded that estimates for some target variables were limited from this type of sampling compromise. However, this has merely manifested in a lack of precision in estimation. As all project partners were informed about this, it can be supposed that an interpretation of these measures was conducted with necessary caution.

Sampling type: We discussed two different approaches on the basis of which one can calculate the sample size (farm or animal as statistical unit). If one were to choose the animal as the statistical unit, one could then make a size-proportional selection of the farms. In this approach, large farms would have a higher probability of being studied than smaller ones because there are more animals there. This approach may initially seem more attractive for the study, as it requires fewer small farms to be studied than the approach on a farm level. This type of selection, however, has far-reaching consequences for subsequent statistical analysis. For descriptive analyses, each farm must be weighted according to its selection probability [17]. For inductive analyses, in which the possible factors influencing the target variables are examined, there are as yet, in some cases, no analysis methods at all with which this weighting can be considered. In particular, if a target variable (i.e., a health outcome) is categorical and, consequently, a logistic regression would have to be performed, there are no analysis methods available so far for this. Overall, therefore, this second type of selection would shift the effort from the phase of investigation to the phase of evaluation, and it would prevent evaluation in parts. Hence, we decided on a multistage sampling of farms and animals within farms. For us, this was the best compromise between good epidemiologic practice, on the one hand, and the comprehensibility of results and feasibility for later use by interested scientists or other parties, on the other hand.

Impact of region: Whether and what kind of regional differences occurred were not the main focuses of this study. Rather, the aim was to investigate a sufficient number of farms for each region. Our decision to divide the study into three regional studies had several advantages, but also disadvantages. It allowed us to adapt the definition of small, medium-sized, and large farms to be region-specific and thus to ensure that all farm sizes were covered by the study sample. However, the general population differed widely between the regions, from only 2256 farms in the region East to more than 32,000 farms in the South. For this reason, the percentage of farms represented in the study varied, leading to a different precision of the estimates. These structural differences also explain why it was possible to study 15% of the target population in the region East with 252 farms, whereas only 2% and 1%, respectively, could be covered with the same number of farms in the regions North and South.

For the estimation of the prevalence rates, this was not a big problem because the number of 250 farms was large enough even for the South to ensure a precision of ± 6% in the case of a 50% prevalence. For regression models and other statistical applications, this discrepancy became important because the percentage of participating farms in the East was higher than in the North or South, and thus the estimation was more precise. Even stratification does not solve this problem, so each model was run separately for each region [17]. In addition, one of the study's objectives was to provide guidance to farmers concerning improvements in animal health and welfare. Due to structural differences, these recommendations needed to be region-specific, and thus data from each region needed to be reliable.

4.2. Representativeness of the Study Population

Source population: To ensure representativeness in terms of farm size, it was important to ensure that the composition of the farms studied reflected the composition of the target population as closely as possible. This was achieved very well in terms of the participating federal states. However, the average size of the PraeRi farms was higher than the data provided by the Stat. Bundesamt (region North: +17%; East: +35%; South: +19%) [18]. In order to keep this effect as low as possible, a sampling plan ensured that 1/3 of the farms per region were small. To classify this apparent discrepancy, it must be pointed out here that—as we described earlier—the farm sizes of dairy cow farms were biased downward in some databases. It should also be noted that the most recent data from the Stat. Bundesamt were from 2016, while our data collection took place until 2019. From this time, it is known that the average farm size had noticeably increased in all regions of Germany due to closures of small farms as well as restocking in the remaining farms (Landwirtschaftskammer Niedersachsen 2019; LKV Berlin-Brandenburg 2020). Comparing the PraeRi study with current reports of the DHI or similar, the PraeRi study represented the composition of the target population in terms of farm sizes well to very well. It can only be speculated as to whether the study results are transferable to other countries or not. We believe that if herd size, husbandry conditions (grazing, stable type), breed, etc., are considered, and also the climate is comparable as, e.g., in the neighboring countries, the study results will be valid to a certain degree.

Already during the preparation of the address data extraction from HIT it became clear that a pure separation of farms keeping dairy cows and other cattle farmers could not be ensured. This is because the information on whether a farm keeps dairy cows is only optional in HIT. For this reason, all cattle farmers who certainly or possibly kept dairy cows were defined as the source population. Before drawing the random selection, it was not possible to determine exactly what proportion of cattle keepers without dairy cows was in the selected sample. Thus, verification of the representative composition of the study population based on farm sizes was not possible without error. In addition to the mixing with suckler cow farms, it also had to be noted that some data in HIT were outdated and HIT listed farms that no longer existed. Since this apparently affected small farms more often than large ones, it can be assumed that the initially selected farm size cut-offs were somewhat too low. This example highlights that even when using an official register, such as the German HIT, data may be outdated or missing, and even mistakes must be taken into consideration.

As the sampling population contained cattle farmers without dairy cows, it must be assumed that the cut-offs determined to balance the sampling plan for farm size were also biased (assumption: the average farm size was slightly underestimated). In addition, the consequence that the source populations from which potential participants were sampled differed between the regions must be regarded as a possible source of selection bias.

Participation rate: The unexpectedly low participation rate required measures to meet the necessary sample size and to cover all categories of the sampling plan sufficiently.

A telephone non-response analysis in region North (n = 20 farms) confirmed that 14 of the farms surveyed did not keep dairy cows or that they planned to give or already had given up the farm. Measures such as tracking and interviewing owners of small farms as well as explicitly addressing small farms in the following letters, both seemed to have had an effect.

After initial difficulties in recruiting small (or, in Bavaria, medium-sized) farms in particular, a separate call for small farms was enclosed with letters of resampling. The importance of the participation of small farms for the significance of the study, and thus the livestock farmers themselves, was emphasized. This improved the participation rate.

The second draw of address data including the adjustment of the farm size categories yielded a full coverage of the sampling plan. Nevertheless, the overall participation rate was very low, at <10% across all regions and all farm size categories, which was much lower than in previous studies. Reasons for this may well be due to a heavy workload in agriculture rather than the scale of data collection in this study, as farmers made the decision to not participate at a time when they were not yet aware of the extent of data collection. In addition, studies with a focus on a specific disease or management problem might be more interesting compared to this rather general topic.

4.3. Prevalence Estimation

The aim of the study was to determine representative and reliable prevalence estimates for the most important dairy cow health parameters. This is because, to our knowledge, such large, time- and cost-consuming studies have been conducted very rarely, if ever, up to now. Alongside this, we were able to provide a data set that allowed for risk factor analyses, as well. The high completeness of the data set can be attributed to the efforts that have been taken by the plausibility checks and the following correction of implausible values whenever possible. Missing values are always a challenge for data analyses and the quality of a data set is also determined by the number of missing values.

This study is an epidemiological study, not an experiment. The study type is referred to as a cross-sectional study and follows specific rules for planning, implementation, and evaluation. It differs from other epidemiological studies in that, among other things, no control group is used here, as no scientifically selected groups are compared, but the status quo in the existing population is surveyed.

A good estimate of disease prevalence rates forms the basis for many studies—observational as well as experimental ones. This information is needed for the calculation of an adequate sample size in future studies that, e.g., want to investigate causal associations. Sample size calculation is necessary in the context of animal experiments. A good knowledge of inner-herd prevalence rates is also necessary to assume the potential improvement of a specific intervention measure. Thus, reliable estimates of prevalence rates in populations are a very important piece of information for current and future research and can thus not be underestimated [19].

Due to the higher number of observers and the fact that they worked at three different universities with possibly different opinions regarding the scoring and severity of lesions and health alterations, three interobserver reliability checks were performed throughout the period of data collection. By combining teaching, discussion, and comparison between the observers during a specifically organized observer meeting, we intended to minimize possible observer bias. Also, the intensive observer seminar at the beginning of data collection as well as the comprehensive SOP catalogue were aimed at minimizing observer bias. The agreement between the observers with regard to the compared characteristics was basically quite good to very good. However, the comparisons could not be carried out with the desirable sample size, as either animal welfare aspects (such as too long fixation of the examined animals) or a lack of available animals for the comparison limited the possible sample size at the selected locations. Here, too, it can be seen that the feasibility of such a large study can stand in the way of desirable methodological quality.

Disease prevalences (Table 6): The right-skewed distribution of the prevalence estimates by the farmers indicates that there were either actually some farms with very high prevalences or that some farmers were able to make a more accurate and thus higher estimation of the prevalences than others. We not only collected the estimation of the farmers in the project, but also made a prevalence determination ourselves. Jensen et al. [8] could show that "on average, farmers were conscious of only 45.3% (North), 24.0% (East), and 30.0% (South) of their lame cows". This fits with the findings of other studies such as Ranjbar et al. [20] who reported a 3.7 times underestimation of lameness in pasture-based herds in Australia. Denis-Robichaud [21] also reported an underestimation of lameness prevalence by farmers and veterinarians. It can be assumed that all farmers' data on disease prevalence are underestimated—especially if the farmers cannot consult any documentation (such as veterinary invoices or DHI reports, etc.) for estimation. Furthermore, regional differences in disease incidence need not necessarily be explained by differences in husbandry or management. Rather, it can be assumed that the differences in the stringency of disease documentation are mainly due to farm size.

Lameness prevalence in our study was a bit higher than in other studies, as recently reported in a literature review on lameness in dairy cows worldwide and over 30 years with study means ranging from 5.1 to 45% [22]. In Europe, only a few recent reports of mastitis prevalence or reproduction parameters exist. One group carried out a meta-analysis on subclinical ketosis and reported a mean global prevalence of 22.7%, which is much higher than in our study [23].

The differences in farm size between the regions resulted in many structural differences, starting with more part-time farmers and less employees including a lower educational level in the South, and more automated milking systems and IVHM contracts in the East. Grazing was most common in the North because farms have more grassland there. The fact that some diseases had lower incidences in the South although the husbandry conditions were not better (e.g., tethering stalls) underpins the multifactorial character of the diseases that were investigated.

In conclusion, we succeeded in having a representative sampling, as much as possible, in terms of herd size and region. Every study with voluntary participation will not be fully representative, and even if it is possible to control for bias in the data analysis (e.g., if the distribution of the herd size in the target population is known), information about underrepresented strata (e.g., small herds) will remain less reliable. In our study, we were able to stratify the farms by herd size and region because we had access to lists of all dairy herds in Germany. If such population data are not available, convenience sampling may also lead to reliable results, as long as the sources of bias are discussed thoroughly.

Number of lactations: Assuming that a cow gave birth to a calf for the first time at the age of 24–36 months, followed by one calf, i.e., one lactation, approximately, per year, it can be estimated that cows reached an age of about four to six years on average. From an economic standpoint, this is not ideal since the full potential of a cow's performance is not reached before the sixth lactation, and rearing costs on average are not paid back earlier than in the third lactation [24]. The differences observed between the regions might be associated with the predominant breed in the regions because the Simmental breed was mainly present in region South and is known to reach higher lactation numbers due to less intensive use.

Culling and Mortality: The term culling includes animals exiting the farm due to slaughter or sale [25]. The reported numbers for culling risk in this study resemble those found in Germany during the last decades without clear changes over the years (BRS 2023). In a meta-analysis by Compton et al. [26], including data from studies in North America, Europe (except Germany) and Australia/New Zealand, lower culling risks were reported (incidence risk from 0.14 to 0.28). The authors also did not find any evidence of overall change in culling incidence risk over time. With regard to mortality, a threshold of 2% has been recommended [27]. In regions North and East, this threshold was exceeded by at least every second farm. In the South, this was the case for at least every fourth farm.

Our figures for mortality rate are in alignment with an earlier study in Germany, where an overall mortality rate of 0.047 per animal-year was reported [28], and other international studies [26]. Due to our study design with the limitation of the observation period to one year, no inference about temporal changes could be drawn. However, in the literature, an increase in mortality has been described [26]. When looking at the reasons as to why animals leave the farm, our data revealed that roughly 50% were due to disease. Due to the fact that two different databases had to be consulted for the analysis of culling and mortality (HIT) and culling reasons (DHI), respectively, there is no information as to why animals die on the farm. A study by Alvåsen et al. [29] showed that the main reasons for mortality and slaughtering differed; e.g., more cows with fertility disorders or udder diseases were slaughtered, whereas metabolic diseases and claw/leg disorders more often were reported for mortality disposal. Accurate documentation, on a farm level, produces information that might be helpful for further disease management.

4.4. Risk Factor Models

Selection and confounding bias need to be considered in observational studies [17]. The way to deal with it is to detect and control it. This is best carried out by drawing causal directional acyclic graphs (causal DAGs) that include not only the influence factor and the outcome of interest, but also all factors that might have led to biased selection and that are a common cause of the influence factor and the outcome. These types of bias can be controlled, e.g., by stratification [30].

As described above, the amount of regression models in the final report was too large to give each of the models the time and effort needed for this well-founded confounder analysis. We needed a more pragmatic approach to be able to finish the report on time. Proper confounder analyses are planned for future scientific publications.

Statistical analysis: Although this was not at all foreseen in the sample design, it was still possible to create risk models with little, describable (and thus controlled) bias. It has also to be stated that the sample size in some cases was not large enough to satisfy the requirements of a risk factor analysis, neither for univariable nor much less for multivariable models.

Our pragmatic approach for the final report followed a standardized and validated procedure that is described above. This procedure was developed and agreed on by all study personnel that conducted regression models. This avoided "statistician-bias" and ensured the highest possible quality and thus can be regarded as a good compromise between scientific demands and resources.

The risk factor models allowed insights into causal associations concerning, e.g., lameness [31], udder health [32], and infectious diseases [33], but also concerning calf and young cattle health [8]. Interestingly, differences could be seen between regions in most cases. These can be used as starting points for future investigations of specific topics.

5. Conclusions

The perfect epidemiologic study does not exist. Non-biased results can only be obtained from experimental studies, and these lack the variance that exists in real life. Therefore, the best that can be achieved is to carry out observational studies that are as representative and as free from bias as possible to record all known sources of confounding and to analyze then and interpret the data as carefully as possible. Not only in large studies but in all, proper project management and comprehensive team communication are necessary to identify and solve all the challenges of a technical or methodic nature that may occur during a study. In the case of the PraeRi study, we have almost representative estimates of the most important animal health disorders in dairy husbandry in Germany. These, in parts alarmingly high, prevalences and incidences, e.g., for lameness, can and have already been used as a basis for in-depth analyses of certain aspects and can contribute a lot to increasing dairy health and welfare.

Supplementary Materials: The following supporting information can be downloaded at https://www.mdpi.com/article/10.3390/ani14091385/s1, S1: Areas for which data were collected in a prevalence study among 765 German dairy herds; S2: Feedback letter to farmers in a prevalence study among 765 German dairy herds; S3: Benchmarking feedback to farmers in a prevalence study among 765 German dairy herds.

Author Contributions: Conceptualization, A.C., R.M., M.H., G.K.-S., M.M., and K.-E.M.; methodology, A.C. and R.M.; validation, A.C., R.M., M.H., G.K.-S., M.M., and K.-E.M.; formal analysis, R.M.; investigation, A.C. and R.M.; writing—original draft preparation, R.M. and A.C.; writing—review and editing, A.C., R.M., M.H., G.K.-S., M.M., and K.-E.M.; visualization, R.M. and A.C.; supervision, A.C., R.M., M.H., G.K.-S., M.M., and K.-E.M.; project administration, A.C., R.M., M.H., G.K.-S., M.M., and K.-E.M.; funding acquisition, A.C., R.M., M.H., G.K.-S., M.M., and K.-E.M. All authors have read and agreed to the published version of the manuscript.

Funding: This research was funded by the German Federal Ministry of Nutrition and Agriculture, grant numbers 2814HS006 (Hannover), 2814HS007 (Berlin), and 2814HS008 (Munich).

Institutional Review Board Statement: Ethical review and approval were waived for the study on human participants in accordance with the local legislation and institutional requirements in 2015. The participants provided their written informed consent to participate in this study. Ethical review and approval were not required for the animal study because no painful interventions were made. This was in accordance with the local legislation and institutional requirements. Written informed consent was obtained from the owners for the participation of their animals in this study.

Informed Consent Statement: Informed consent was obtained from all farmers involved in the study.

Data Availability Statement: The data sets presented in this article are not readily available because the data were acquired through cooperation between different universities. Therefore, any data transfer to interested persons is not allowed without an additional formal contract. Data are available for qualified researchers who sign a contract with the project consortium. This contract will include guarantees of the obligation to maintain data confidentiality in accordance with the provisions of German data protection law. Currently, there exists no data access committee nor another body who can be contacted for the data; a committee will be founded for this purpose. This future committee will consist of the authors as well as members of the related universities. Interested cooperative partners, who are able to sign a contract as described above, may contact MH, Clinic for Cattle at the University of Veterinary Medicine, Hannover, Bischofsholer Damm 15, 30173 Hannover, Germany, Email: martina.hoedemaker@tiho-hannover.de. Requests to access the data sets should be directed to martina.hoedemaker@tiho-hannover.de.

Acknowledgments: The authors wish to thank all participating dairy farmers and staff for their cooperation, and the students and research assistants for their support. Furthermore, the authors wish to thank all persons involved in realizing the PraeRi project: Friedemann Adler, Heidi Arndt, Alexander Bartel, Julia Bäumer, Katrin Birnstiel, Ann-Kathrin Bodenstein, Alexander Choucair, Linda Dachrodt, Marcus Doherr, Phuong Do Duc, Melanie Feist, Natascha Gundling, Antonia Hentzsch, Miriam Hielscher, Bernadette Hinzmann, Katharina Charlotte Jensen, Verena Kaufmann, Laura Kellermann, Marcus Klawitter, Lothar Kreienbrock, Corinna Lausch, Iris Litjens, Rolf Mansfeld, Kim Meier, Andreas Öhm, Philip Paul, Frederike Reichmann, Anne Rösler, Dmitrij Sartison, Annegret Stock, Alexander Stoll, Maria Volkmann, Marina Volland, Svenja Woudstra, and Philipp Zuz.

Conflicts of Interest: The authors declare no conflicts of interest.

References

1. Hoffmann, W.; Latza, U.; Baumeister, S.E.; Brünger, M.; Buttmann-Schweiger, N.; Hardt, J.; Hoffmann, V.; Karch, A.; Richter, A.; Schmidt, C.O.; et al. Guidelines and recommendations for ensuring Good Epidemiological Practice (GEP): A guideline developed by the German Society for Epidemiology. *Eur. J. Epidemiol.* **2019**, *34*, 301–317. [CrossRef] [PubMed]
2. Sargeant, J.M.; O'Connor, A.M.; Dohoo, I.R.; Erb, H.N.; Cevallos, M.; Egger, M.; Ersbøll, A.K.; Martin, S.W.; Nielsen, L.R.; Pearl, D.L.; et al. Methods and processes of developing the strengthening the reporting of observational studies in epidemiology – veterinary (STROBE-Vet) statement. *Prev. Vet. Med.* **2016**, *134*, 188–196. [CrossRef] [PubMed]
3. Merle, R.; Busse, M.; Rechter, G.; Meer, U. Regionalisierung Deutschlands anhand landwirtschaftlicher Strukturdaten: Regionalisation of Germany by data of agricultural structures. *Berl. Munch. Tierärztl. Wochenschr.* **2012**, *125*, 52–59. [PubMed]
4. Glaser, S.; Kreienbrock, L. *Stichprobenplanung bei veterinärmedizinischen Studien: Ein Leitfaden zur Bestimmung des Untersuchungsumfangs*, 1st ed.; Schlütersche: Hannover, Germany, 2011; ISBN 978-3-89993-078-8.

5. Bayrisches Staatsministerium für Ernährung, Land- und Forstwirtschaft. Herkunftssicherungs- und Informationssystem für Tiere. Available online: https://www.hi-tier.de/default.htm (accessed on 21 August 2023).
6. Jensen, K.C.; Frömke, C.; Schneider, B.; Do Duc, P.; Gundling, F.; Birnstiel, K.; Schönherr, F.; Scheu, T.; Kaiser-Wichern, A.; Woudstra, S.; et al. Case-control study on factors associated with a decreased milk yield and a depressed health status of dairy herds in northern Germany. *BMC Vet. Res.* **2019**, *15*, 442. [CrossRef] [PubMed]
7. State Control Association Berlin-Brandenburg e.V. Annual Report 2019. Available online: https://www.lkvbb.de/fileadmin/Redaktion/Publikationen/2020/Bericht/Jahresbericht-2019-Onlineausgabe.pdf (accessed on 31 August 2023).
8. Dachrodt, L.; Arndt, H.; Bartel, A.; Kellermann, L.M.; Tautenhahn, A.; Volkmann, M.; Birnstiel, K.; Do Duc, P.; Hentzsch, A.; Jensen, K.C.; et al. Prevalence of disorders in preweaned dairy calves from 731 dairies in Germany: A cross-sectional study. *J. Dairy Sci.* **2021**, *104*, 9037–9051. [CrossRef] [PubMed]
9. Edmonson, A.J.; Lean, I.J.; Weaver, L.D.; Farver, T.; Webster, G. A Body Condition Scoring Chart for Holstein Dairy Cows. *J. Dairy Sci.* **1989**, *72*, 68–78. [CrossRef]
10. Sprecher, D.J.; Hostetler, D.E.; Kaneene, J.B. A lameness scoring system that uses posture and gait to predict dairy cattle reproductive performance. *Theriogenology* **1997**, *47*, 1179–1187. [CrossRef] [PubMed]
11. Leach, K.A.; Dippel, S.; Huber, J.; March, S.; Winckler, C.; Whay, H.R. Assessing lameness in cows kept in tie-stalls. *J. Dairy Sci.* **2009**, *92*, 1567–1574. [CrossRef]
12. Cook, N.B.; Reinemann, D. *A Tool Box for Assessing Cow, Udder and Teat Hygiene*; University of Wisconsin-Madison: Madison, Wisconsin, 2007.
13. Kellermann, L.M.; Rieger, A.; Knubben-Schweizer, G.; Metzner, M. Short communication: Design and validation of a hygiene score for calves. *J. Dairy Sci.* **2020**, *103*, 3622–3627. [CrossRef]
14. Lower Saxony Chamber of Agriculture. *Performance and Quality Audits and Projects in Animal Husbandry—Annual Report 2018*; Willers Druck GmbH & Co KG: Oldenburg, Germany, 2020; ISBN 978-3-9818882-2-5.
15. Ruddat, I.; Scholz, B.; Bergmann, S.; Buehring, A.-L.; Fischer, S.; Manton, A.; Prengel, D.; Rauch, E.; Steiner, S.; Wiedmann, S.; et al. Statistical tools to improve assessing agreement between several observers. *Animal* **2014**, *8*, 643–649. [CrossRef]
16. Textor, J.; van der Zander, B.; Gilthorpe, M.S.; Liśkiewicz, M.; Ellison, G.T.H. Robust causal inference using directed acyclic graphs: The R package 'dagitty'. *Int. J. Epidemiol.* **2016**, *45*, 1887–1894. [CrossRef] [PubMed]
17. Dohoo, I.R.; Martin, W.; Stryhn, H. *Methods in Epidemiologic Research*; VER: Charlottetown, PE, Canada, 2012; ISBN 978-0-919013-73-5.
18. Statistisches Bundesamt. Land- und Forstwirtschaft, Fischerei: Viehbestand. Fachserie 3 Reihe 4.1 [Excel-Table]; Wiesbaden, 2014 (Artikelnummer: 2030410145315). Available online: https://www.destatis.de/DE/Themen/Branchen-Unternehmen/Landwirtschaft-Forstwirtschaft-Fischerei/Tiere-Tierische-Erzeugung/Publikationen/Downloads-Tiere-und-tierische-Erzeugung/viehbestand-2030410225324.html (accessed on 15 October 2019).
19. Thrusfield, M.; Ortega, C.; de Blas, I.; Noordhuizen, J.P.; Frankena, K. WIN EPISCOPE 2.0: Improved epidemiological software for veterinary medicine. *Vet. Rec.* **2001**, *148*, 567–572. [CrossRef]
20. Ranjbar, S.; Rabiee, A.R.; Ingenhoff, L.; House, J.K. Farmers' perceptions and approaches to detection, treatment and prevention of lameness in pasture-based dairy herds in New South Wales, Australia. *Aust. Vet. J.* **2020**, *6*, 264–269. [CrossRef]
21. Denis-Raubichaud, J.; Kelton, D.; Fauteux, V.; Villettaz-Robichaud, M.; Dubuc, J. Short communication: Accuracy of estimation of lameness, injury, and cleanliness prevalence by dairy farmers and veterinarians. *J. Dai. Sci.* **2020**, *103*, 10696–10702. [CrossRef] [PubMed]
22. Thomsen, P.T.; Shearer, J.K.; Houe, H. Prevalence of lameness in dairy cows: A literature review. *Vet. J.* **2023**, *295*, 105975. [CrossRef]
23. Loiklung, C.; Sukon, P.; Thamrongyoswittayakul, C. Global prevalence of subclinical ketosis in dairy cows: A systematic review and meta-analysis. *Res. Vet. Sci.* **2022**, *144*, 66–76. [CrossRef]
24. Boulton, A.C.; Rushton, J.; Wathes, D.C. An empirical analysis of the cost of rearing dairy heifers from birth to first calving and the time taken to repay these costs. *Animal* **2017**, *11*, 1372–1380. [CrossRef] [PubMed]
25. Fetrow, J.; Nordlund, K.V.; Norman, H.D. Invited review: Culling: Nomenclature, definitions, and recommendations. *J. Dairy Sci.* **2006**, *89*, 1896–1905. [CrossRef] [PubMed]
26. Compton, C.W.R.; Heuer, C.; Thomsen, P.T.; Carpenter, T.E.; Phyn, C.V.C.; McDougall, S. Invited review: A systematic literature review and meta-analysis of mortality and culling in dairy cattle. *J. Dairy Sci.* **2017**, *100*, 1–16. [CrossRef]
27. Kuratorium für Technik und Bauwesen in der Landwirtschaft e.V. KTBL-Jahresbericht 2020, Darmstadt, 2021. Available online: https://www.ktbl.de/fileadmin/user_upload/Allgemeines/Jahresbericht/2020/KTBL-Jahresbericht_2020.pdf (accessed on 17 February 2024).
28. Pannwitz, G. Standardized analysis of German cattle mortality using national register data. *Prev. Vet. Med.* **2015**, *118*, 260–270. [CrossRef]
29. Alvåsen, K.; Jansson Mörk, M.; Dohoo, I.R.; Sandgren, C.H.; Thomsen, P.T.; Emanuelson, U. Risk factors associated with on-farm mortality in Swedish dairy cows. *Prev. Vet. Med.* **2014**, *117*, 110–120. [CrossRef] [PubMed]
30. Hernán, M.A.; Robins, J.M. *Causal Inference: What If*; Chapman & Hall/CRC: Boca Raton, FL, USA, 2020.

31. Oehm, A.W.; Merle, R.; Tautenhahn, A.; Jensen, K.C.; Mueller, K.-E.; Feist, M.; Zablotski, Y. Identifying cow—Level factors and farm characteristics associated with locomotion scores in dairy cows using cumulative link mixed models. *PLoS ONE* **2022**, *17*, e0263294. [CrossRef] [PubMed]
32. Böker, A.R.; Bartel, A.; Do Duc, P.; Hentzsch, A.; Reichmann, F.; Merle, R.; Arndt, H.; Dachrodt, L.; Woudstra, S.; Hoedemaker, M. Status of udder health performance indicators and implementation of on farm monitoring on German dairy cow farms: Results from a large scale cross-sectional study. *Front. Vet. Sci.* **2023**, *10*, 1193301. [CrossRef] [PubMed]
33. Abele, G.E.; Zablotski, Y.; Feist, M.; Jensen, K.C.; Stock, A.; Campe, A.; Merle, R.; Oehm, A.W. Prevalence of and factors associated with swellings of the ribs in tie stall housed dairy cows in Germany. *PLoS ONE* **2022**, *17*, e0269726. [CrossRef] [PubMed]

animals

MDPI

Article

Epidemiology Tools to Evaluate the Control of Proliferative Enteropathy in Commercial Pig Herds

Alison Collins [1,*] and Cherie Collins [2]

1 New South Wales Department of Primary Industries, Elizabeth Macarthur Agricultural Institute, Menangle, NSW 2568, Australia
2 Rivalea Australia, P.O. Box 78, Corowa, NSW 2646, Australia; ccollins@rivalea.com.au
* Correspondence: alison.collins@dpie.nsw.gov.au

Simple Summary: Evaluating disease control in commercial pig herds requires the use of infection surveillance tools as well as observing clinical signs and measuring production changes. Many disease control studies evaluate treatments in research facilities where conditions do not equate with commercial farms, specifically with respect to the presence of multiple and repeated challenges to pigs. This study evaluated the control of the wasting disease proliferative enteropathy with disinfection, vaccination, and/or medication in 84 pig pens under commercial conditions using epidemiological tools. Vaccinated pigs housed in lime-disinfected pens showed the best control of proliferative enteropathy with reduced diarrhea and fecal shedding of *Lawsonia intracellularis* and an increased growth rate and prevalence of protective antibodies. Continuous medication with olaquindox prevented *L. intracellularis* infection until it was removed at 17 weeks of age, leaving naïve finisher pigs susceptible to intestinal hemorrhage, bloody feces, reduced growth, and mortalities. To avoid proliferative enteropathy, control strategies need to suppress but not prevent *L. intracellularis* infection, while enabling protective antibodies to develop. This study demonstrated the value of monitoring the timing and level of excretion of *L. intracellularis* in pig pens to compare proliferative enteropathy control on commercial farms.

Abstract: Proliferative enteropathy (PE) is characterized by diarrhea and reduced weight gains in growing pigs and intestinal hemorrhage in finishers. Vaccination, antibiotic medication, and improved hygiene can control PE, but their efficacy depends upon the epidemiology of PE. This study monitored the timing and severity of PE in 84 commercial pens across seven treatments, including disinfection, vaccination, no treatment, medication with olaquindox (50, 25 and 12.5 ppm), and combined disinfection and vaccination. Vaccination with or without lime disinfection suppressed clinical signs of PE and reduced the number of excreted *L. intracellularis* relative to untreated pigs housed in cleaned or cleaned and disinfected pens between 9 and 17 weeks of age. Continuous olaquindox mediation to 17 weeks of age prevented *L. intracellularis* infection, leaving finisher pigs naïve. These finisher pigs suffered an outbreak of hemorrhagic enteropathy with significant reductions in weight gain, feed intake, and mortalities of 4.6%. Over the 13 week grow/finish period, vaccinated pigs housed in disinfected pens showed significantly higher weight gain and feed intake relative to all other treatments, equating to a weight gain difference of between 3.6 and 3.9 kg per pig. Monitoring the immune response and fecal excretion of *L. intracellularis* in pens of pigs enabled effective PE control strategies to be evaluated on the farm.

Keywords: proliferative enteropathy (PE); *Lawsonia intracellularis*; disinfection; vaccination; seroconversion; diarrhea; hemorrhage; olaquindox; epidemiology

1. Introduction

Proliferative enteropathy (PE) is a wasting disease of grower and finisher pigs caused by *Lawsonia intracellularis*, an intracellular bacterium that replicates in the cytoplasm of epithelial cells lining the ileum and causes thickening of the mucosa due to proliferation of immature enterocytes [1,2]. The clinical signs of PE depend on the dose of *L. intracellularis* [3–5] and the age and susceptibility of pigs, ranging from diarrhea and reduced growth rates in grower pigs to acute hemorrhagic enteropathy and mortality in up to 10% of affected finisher pigs [6,7]. Economic losses associated with clinical and sub-clinical PE have been estimated at between AUD 8 and AUD 13 per pig (between USD 5 and USD 8.30) in reduced net revenue [8]. Losses are due to reduced weight gains, variation in pig weights within batches, poor feed efficiency, mortalities, and increased days to slaughter.

Serological surveys have demonstrated between 80% and 100% of pig herds are infected with *L. intracellularis* [9,10] with an average of 84% of finisher pigs infected within herds. Pigs are susceptible to *L. intracellularis* after weaning, with the loss of maternal antibodies provided in dams' milk [11]. However, the common inclusion of antibiotics in weaner diets to control other enteric or respiratory diseases may prevent *L. intracellularis* infection until pigs are moved to grower or finisher accommodation. Serological surveys on commercial pig farms demonstrate an increased seroprevalence to *L. intracellularis* in older animals [10,12–15]. *L. intracellularis* is transmitted between pigs by the fecal-oral route [4,16,17], so poor pen hygiene, continuous flow production, an accumulation of manure, and the mixing of pigs increases the risk of *L. intracellularis* infection [13,18,19]. Natural exposure to *L. intracellularis* and vaccination induce protective immunity to *L. intracellularis*, including specific antibodies (IgA and IgG) and cell-mediated (interferon gamma) immune responses in serum and intestinal mucosa [4,20–23].

Experimental infection studies have demonstrated that vaccination [20] and antibiotic medication [24–29] can reduce clinical signs of PE, reduce fecal excretion of *L. intracellularis* and reduce histopathologic lesions of PE in affected pigs. However, many PE control studies evaluate treatments in research facilities where conditions do not equate with commercial farms, specifically with respect to the presence of multiple and repeated challenges to pigs, stocking density, hygiene, and environmental conditions. In addition, experimental challenge studies that define PE control as the reduced severity of histopathology lesions require necropsy of pigs at the peak of intestinal pathology and may thus underestimate the production losses associated with a more protracted recovery.

This study aimed to demonstrate the value of molecular and serological epidemiological tools to compare the efficacy of PE control strategies commonly used on commercial pig farms, noting that fecal excretion of *L. intracellularis* has been proven to be correlated with the severity of PE lesions and clinical signs of PE [30–33]. PE control strategies to be evaluated include the published treatments of olaquindox [29,34], vaccination with Enterisol® Ileitis [20], disinfection [35], and tylosin [25].

2. Materials and Methods

2.1. Animals, Housing, and Experimental Design

The study was conducted according to the Australian Code for the Care and Use of Animals for Scientific Purposes, and animal ethics approval was granted by the Animal Ethics Committees of both the commercial pig production company and the Elizabeth Macarthur Agricultural Institute (13V057C). Approximately 3024 grower pigs (9-week-old females and males) were housed in 84 pens (42 male and 42 female) in a high health status commercial piggery in south-eastern Australia. Each pen was allocated to one of seven treatments (Table 1). All pens were pressure washed with water prior to filling over two weeks with three replicates each week, equaling a total of six replicates. Pens had a mix of solid and concrete slatted floors (33:66) with mesh dividers between each pen. Approximately 35 $\pm$ 2 pigs were housed in each pen and pigs had ad libitum access to feed and water. The base grower diet was comprised mainly of wheat, canola meal, meat meal, and tallow and was formulated to meet the animals' requirements, including 10.1 MJ net energy per kg, 17.7% crude protein, 2.97% fat, 48.35% starch, 3.8% fiber, and 1.12% lysine. Finisher base diets were primarily composed of wheat, barley, canola meal, meat meal, and tallow and were formulated to 9.94 MJ net energy per kg, 14.2% crude protein, 1.96% fat, 51.55% starch, 3.75% fiber, and 0.87% lysine.

Table 1. Treatments and number of pens per treatment.

Treatment Group	Number of Pens	Treatment Description
A	14	Lime disinfectant + vaccination
B	14	Vaccination
C	12	Lime disinfectant
D	8	No treatment
E	14	12 ppm Olaquindox
F	14	25 ppm Olaquindox
G	8	50 ppm Olaquindox
Total	84	

All pigs were medicated with 50 ppm olaquindox (Keyquindox, International Animal Health Products, Huntingwood, NSW, Australia) from 3 to 9 weeks of age, prior to the start of the trial. Pigs in treatment groups A and B were vaccinated orally with Enterisol® Ileitis (Boehringer Ingelheim Animal Health USA Inc., Duluth, GA, USA) between two and three weeks of age, leaving a three-day antibiotic-free window before and after vaccination. Pens for treatments A and C were disinfected with lime (Hydrated lime, plasterer's quality) prior to the grower pigs entering at 9 weeks of age. These lime-disinfected pens were grouped together to avoid cross-contamination with other pens. The pens allocated to 50 ppm of Olaquindox were also grouped together and placed adjacent to the lime-treated pens with solid dividers between treatments. All other treatments were randomly allocated to the remaining pens (Figure 1).

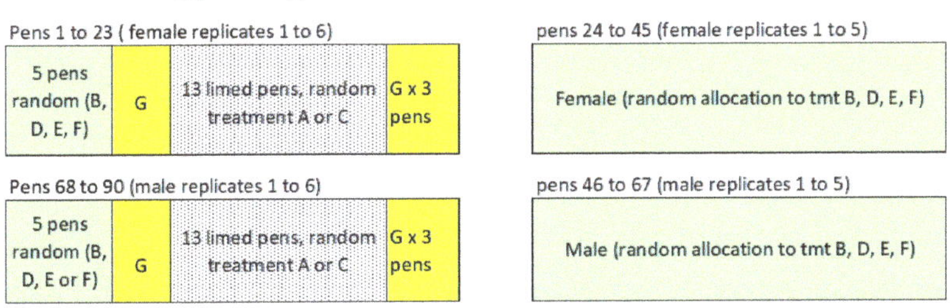

Figure 1. Schematic diagram of pens and treatment layout within sheds.

Olaquindox was maintained in the diets of treatment E, F, and G pigs until day 56 of the trial (16–17 weeks of age) when it was replaced with a two-day pulse of 400 ppm tylosin (Tylan 250, Elanco @ 16 mg/kg bodyweight, Greenfield, IN, USA) in feed every 10 days from trial day 56 until sale at 21–22 weeks of age. Tylan was also pulsed in feed to pigs in treatments A to D from day 57 onwards due to the occurrence of scouring in more than 10% of the pigs in each pen.

As this trial was conducted in a commercial herd, normal health management practices were employed to control other diseases. All pigs were vaccinated against *Actinobacillus pleuropneumoniae* (APP) with an autogenous serovar 15 bacterin at 8, 10, 12, and 16 weeks of age. Pigs were fed a commercial grower diet from 9 to 16 weeks of age, followed by a commercial finisher diet from 16 to 22 weeks of age. All feed contained the organic acid Fysal @ 2 kg/t (Selko Feed Additives, Amersfoort, The Netherlands) for Salmonella control and the antimicrobial salinomycin at 60 ppm (BioCox 120 g/kg, Huvepharma, Sofia, Bulgaria) for control of *Brachyspira* spp. Male pigs were vaccinated twice with Improvac (Zoetis, Parsippany, NJ, USA) at 13 and 19 weeks of age. All trial pigs were selected from dams vaccinated against porcine circovirus (PCV2), *Pasteurella multocida*, APP, Erysipelas, Parvovirus, Leptospirosis, Glassers disease, and *Mycoplasma hyopneumoniae*.

2.2. Production Measures and Analysis

Pigs were weighed, and their feed intake was measured at six time periods (day 0 on entry, day 21, 42, 56, 70, and 91). The pen weights and number of pigs per pen were recorded and the average daily gain (ADG) was calculated by subtracting the earlier pen weight from the later pen weight and dividing by the number of pigs and number of days between the first and second weights. Average daily feed intake (ADFI) was calculated by subtracting the weight of feed not consumed from the total weight of feed provided over each period and dividing this amount by the number of pigs in the pen and the number of days. The feed conversion ratio (FCR) was calculated as the feed intake divided by the ADG. At slaughter (day 91), the liveweight, hot standard carcass weight (HSCW), the P2 backfat and dressing percent were recorded for each pig. The number of intestines condemned at slaughter was only recorded for the fifth and sixth replicates and expressed as a percent of the total pigs.

2.3. Fecal Sampling, Nucleic Acid Extraction, and Quantitative Polymerase Chain Reaction (qPCR)

Two pooled fecal samples (each containing 5 individual samples randomly selected from the pen floor) were collected from each pen at six time points (day 21, 35, 49, 56, 70, 84). Previous studies demonstrated that random sampling of five fecal samples per pen floor provided an accurate measure of *L. intracellularis* numbers [36]. Nucleic acids were extracted from the feces using a MagMax DNA extraction kit (Applied Biosystems, Foster City, CA, USA) as previously described [37]. Nucleic acids from fecal extracts were amplified in a real-time polymerase chain reaction (RT PCR), alongside fecal standards seeded with known numbers of *L. intracellularis* (10^4 to 10^8 *L. intracellularis* per gram of feces). A standard curve from the seeded feces was used to quantify *L. intracellularis* numbers in the pooled pen samples.

2.4. Serum IgG Response to L. intracellularis

Blood from about 11% of the pigs in each pen were collected at three time points (day 30, 63, and 85) to confirm the timing of seroconversion. Serum IgG antibodies to *L. intracellularis* were detected with a commercial competitive (blocking) ELISA kit (Bioscreen® Ileitis, GmbH, Münster, Germany) with a positive index cut-off at 30% inhibition as recommended by the manufacturer.

2.5. Statistical Analysis

The effect of treatment on production and disease measures was investigated using an unbalanced analysis of variance (ANOVA) with replicate, gender, and their interactions as blocking effects and starting weight (day 0 = 9 weeks age) as a covariate. The least significance difference (LSD) t test was used post hoc on ANOVA to determine significant differences between treatments, and the average standard error of difference was noted for each analysis. The same data were also modelled with a restricted maximum likelihood linear mixed model (REML) with fixed effects of treatment, gender, starting weight at day 0, and all interactions. Random effects included replicate and interactions between replicate, treatment, gender, and starting weight. Variance for all interactions was constrained to be positive. Significant differences between means were illustrated in figures using the glmmTMB function [38] in the R statistical software (R Core Team 2024, R-4.3.3). A compact letter display (CLD) for pairwise comparisons between treatments for each sex (at each timepoint) was determined using the cld function in the multcomp package [39] with the default Tukey *p*-value adjustment, applied on the marginal means for treatments (within each sex and timepoint) calculated using the emmeans package [40].

Samples below the detection limit for the *L. intracellularis* qPCR were randomly allocated a number under 1000, the limit of quantification of the qPCR. The number of *L. intracellularis* was $\log_{10}$ transformed to normalize the distribution of data. A Spearman's rank correlation (non-parametric) was used to test the association between *L. intracellularis* numbers and the ADG, ADFI, and FCR (Genstat 16th edn, VSN International 2013). Prior to correlation analysis, the ADG was divided by the starting weight at week 9, to compensate for differences between the six replicates.

3. Results

3.1. *Epidemiology of PE in Pigs 9 to 12 Weeks of Age (Trial Days 0 to 21)*

Pig starting weights at 9 weeks of age (day 0) were not significantly different between genders, but they were significantly different between replicates, with higher weights in replicates 4 to 6, relative to replicates 1 to 3. There was no serological or qPCR evidence of *L. intracellularis* infection in any of the treatments in the first three weeks of the trial (up to day 21). Neither treatment, nor gender, nor replicate had a significant effect on ADG. However, gender, replicate, and starting weight affected the FCR and ADFI in this period. There was some evidence of APP infection in this period, which caused lost production and mortalities that were not significantly different between treatments.

3.2. *Epidemiology of PE in Pigs 12 to 15 Weeks of Age (Trial Days 21 to 42)*

L. intracellularis infection was detected in treatments C (lime) and D (control) at day 35, with these treatments excreting significantly more *L. intracellularis* than treatments A (vaccine + lime), E, F, and G (medicated with 12, 25, and 50 ppm olaquindox, respectively) (Table 2). Control pigs (D) also shed more *L. intracellularis* than pigs in treatment B (vaccine alone). Over this period, pigs across all treatments were clinically affected with APP and swine dysentery, with mortalities ranging from 0.2% to 1.8% of pigs. Swine dysentery accounted for 20 of the 67 deaths, and APP caused 35 of 67 deaths. However, mortalities were not significantly different between treatments.

Table 2. Mean number of *L. intracellularis* excreted (Log_{10}/g feces) at days 35, 49, and 56 for treatments A to G (vaccine plus lime; vaccine only; lime only; no treatment; 12 ppm; 25 ppm; and 50 ppm olaquindox), average standard error of difference (Av SED) and F probability.

Treatment		Log_{10} *L. intracellularis*/g Feces		
Group	**Description**	**Day 35**	**Day 49**	**Day 56**
A	Disinfectant + vaccination	2.631 ^{a}	5.665 ^{b}	8.852 ^{b}
B	Vaccination	2.925 ^{a}	5.111 ^{b}	7.351 ^{b}
C	Disinfectant	3.741 ^{b}	5.136 ^{b}	7.783 ^{b}
D	No treatment	4.091 ^{b}	6.288 ^{b}	8.884 ^{b}
E	12 ppm Olaquindox	2.753 ^{a}	2.702 ^{a}	4.372 ^{a}
F	25 ppm Olaquindox	2.600 ^{a}	2.713 ^{a}	3.068 ^{a}
G	50 ppm Olaquindox	2.482 ^{a}	2.453 ^{a}	3.349 ^{a}
Av SED		0.4482	0.8706	0.9097
ANOVA		$p = 0.005$	$p < 0.001$	$p < 0.001$

Different superscripts within columns indicate significant differences.

ADG was affected by replicate between days 21 and 42, but this may have been a consequence of differences in starting weights, with replicates 4 to 6 heavier than replicates 1 to 3. Although there was no overall treatment effect on ADG, least square differences from the ANOVA indicate that ADG was lower in treatments B (vaccine only) and D (controls) relative to treatments A (vaccine plus lime), C (lime), and G (50 ppm olaquindox) between days 21 and 42. Vaccination alone (B) appeared to have no effect on the ADG, relative to the control, i.e., the untreated pigs. In contrast, vaccination plus lime (A), lime alone (C), and olaquindox treatments (E, F, and G), improved the ADG relative to the control (D) treatment (Figure 2).

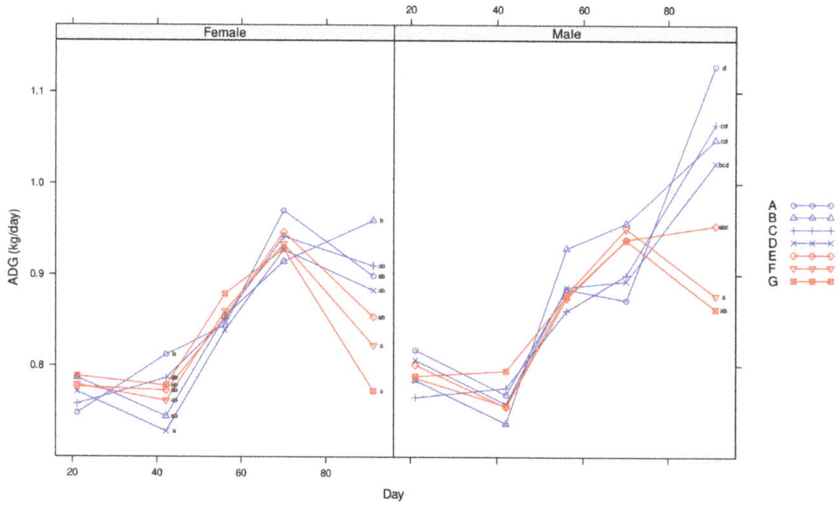

Figure 2. Mean ADG over time for females (**left side**) and males (**right side**) for treatments A to G (Enterisol vaccine plus lime; vaccine only; lime only; no treatment; 12 ppm; 25 ppm; and 50 ppm olaquindox, respectively). Different superscripts indicate significant differences.

Between day 21 and 42, treatment had a significant effect on the ADFI, along with replicate, gender, starting weights, and the interaction between replicate and gender ($p = 0.05$). Vaccinated pigs in lime-treated pens (A) had a higher ADFI than all other treatments (B to F), except treatment G (50 ppm olaquindox). Treatment G pigs had a higher ADFI than the control (D) and vaccinated (B) pigs. Females had a significantly higher ADFI than males (1.69 and 1.59 kg, respectively, $p < 0.001$). Feed intake was also higher in replicates 4 to 6 relative to replicates 1 to 3, which may be explained by higher starting weights in replicates 4 to 6. Feed efficiency was not significantly different between treatments, but both gender and replicate affected the FCR. No significant correlation was observed between the number of *L. intracellularis* excreted (day 21 and 35) and the ADG ($R = -0.067$, $p = 0.545$), ADFI ($R = -0.030$, $p = 0.786$), and FCR ($R = 0.039$, $p = 0.727$) between days 21 and 42.

3.3. Epidemiology of PE in Pigs 15 to 17 Weeks of Age (Trial Days 42 to 56)

Pigs medicated with olaquindox continued to show no evidence of *L. intracellularis* infection until day 56, when *L. intracellularis* were first detected in treatment E (12 ppm olaquindox) feces. Treatments A to D demonstrated a significantly higher excretion of *L. intracellularis* at day 56 and antibodies to *L. intracellularis* at day 63 relative to treatments E, F, and G (medicated with olaquindox) (Table 2 and Figure 3).

Neither the treatment nor starting weight affected the ADG, ADFI, or FCR. However, there was a gender effect on all production measures (Figure 2), and there was a replicate effect on the ADFI and FCR. Although some diarrhea was observed in treatments A to D between days 42 and 56, coinciding with a significantly higher fecal excretion of *L. intracellularis* (Table 2), no significant differences in the ADG, ADI, or FCR were observed in treatments A to D relative to the olaquindox treatments E, F, and G. There was a significant negative correlation between the number of excreted *L. intracellularis* and the ADG (relative to the starting weight) between days 21 and 56 ($R = -0.350$, $p = 0.016$). Minimal mortalities occurred in this period.

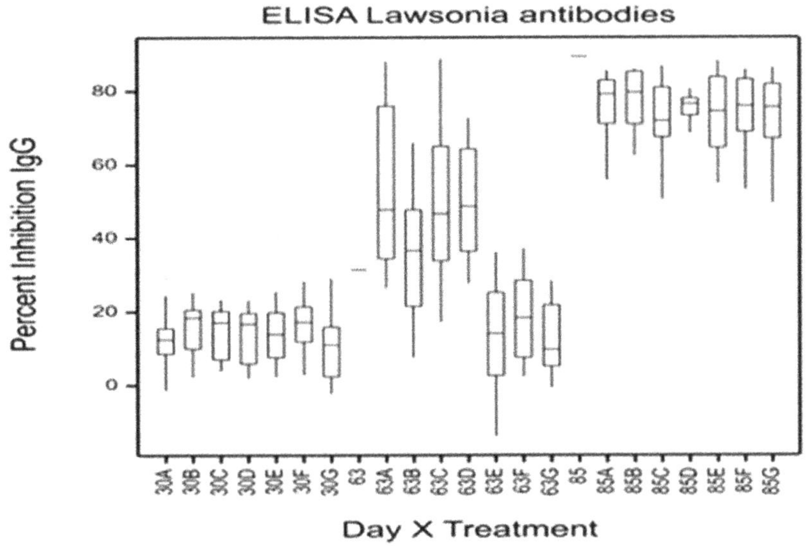

Figure 3. Box and whisker plot of mean serum concentration of IgG antibodies to *L. intracellularis* at day 30, 63, and 85 for treatments A to G (vaccine plus lime; vaccine only; lime only; no treatment; 12 ppm; 25 ppm; and 50 ppm olaquindox). A percent inhibition less than 30% was considered negative. Descriptive statistics were not tested for significance. Green cross indicates outlier value.

3.4. Epidemiology of PE in Pigs 17 to 19 Weeks of Age (Trial Days 56 to 70)

Diarrhea was observed in treatments A to D around 56 days, and these treatments were medicated in feed with tylan from day 57 onwards. At day 70, significantly higher numbers of *L. intracellularis* were detected in treatments A, B, and C relative to both treatments F and G ($p = 0.004$) (Table 3). *L. intracellularis* numbers in treatments D (control) and E (12 ppm olaquindox) were also higher than treatments F and G (25 and 50 ppm olaquindox) (Table 3). However, neither treatment, gender, nor replicate had a significant effect on the ADG, ADFI, or FCR (Figure 2), but the starting weight did significantly affect the ADFI ($p < 0.05$). There was a significant negative correlation between *L. intracellularis* numbers and the relative ADG between days 56 and 70 ($r = -0.580$, $p < 0.001$). Minimal mortalities occurred in this period.

Table 3. Mean number of *L. intracellularis* excreted (Log_{10}/g feces) at days 70 and 84 for treatments A to G (vaccine plus lime; vaccine only; lime only; no treatment; 12 ppm; 25 ppm; and 50 ppm olaquindox), average standard error of difference (Av SED) and F probability.

Treatment		Log_{10} *L. intracellularis*/g Feces	
Group	**Description**	**Day 70**	**Day 84**
A	Disinfectant + vaccination	8.643 [c]	7.314 [bc]
B	Vaccination	8.490 [c]	8.177 [b]
C	Disinfectant	8.837 [c]	7.210b [c]
D	No treatment	8.191 [b]	6.390 [c]
E	12 ppm Olaquindox	7.416 [b]	9.777 [a]
F	25 ppm Olaquindox	6.805 [a]	10.201 [a]
G	50 ppm Olaquindox	5.723 [a]	10.555 [a]
Av SED		0.8081	0.7804
ANOVA		$p = 0.004$	$p < 0.001$

Different superscripts within columns indicate significant differences.

3.5. Epidemiology of PE in Pigs 19 to 21 Weeks of Age (Trial Days 70 to 84)

Bloody diarrhea was observed from day 79 onwards in treatments E, F, and G, which required treatment of individual pigs with injectable Lincomycin (Lincomix RTU, Zoetis, Kalamazoo, MI, USA). Mortalities due to PE and swine dysentery occurred in all treatments, but the highest mortalities occurred in the olaquindox treatments (E and G: 12 and 50 ppm olaquindox), with mortalities of 3.1% and 4.6%, respectively. However, there was no statistically significant difference in mortalities between any treatments over this period ($p = 0.149$) or over the whole trial period (day 0 to 91).

Excretion of *L. intracellularis* continued to be detected in all groups at days 70 and 84. However, at day 84, significantly higher numbers of *L. intracellularis* were detected in treatments E, F, and G (previously medicated with olaquindox) relative to treatments A to D (Table 3). In addition, the control treatment (D) had the lowest mean number of *L. intracellularis*, which was significantly different to treatments B, E, and F ($p < 0.001$). The mean concentration of antibodies to *L. intracellularis* was not significantly different ($p = 0.941$) between treatments at day 85 (Figure 3).

Both treatment and gender had a significant effect on the ADG ($p < 0.001$), with a higher ADG for treatments A, B, and C relative to treatments E, F, and G (Figure 4). Higher ADG was also observed in control pigs (D) relative to pigs previously medicated with olaquindox (treatments F and G) (Figure 4). Treatment, gender and replicate all had a significant effect on the ADFI with a lower ADFI in pigs previously medicated with olaquindox (E, F, and G) relative to treatments A and B. The FCR between treatments was close to significantly different ($p = 0.052$) with a numerically higher FCR in pigs previously medicated with 50 ppm olaquindox (G) compared with pigs in treatments B and C. In this period, gender and replicate also had a significant effect on the FCR ($p < 0.001$).

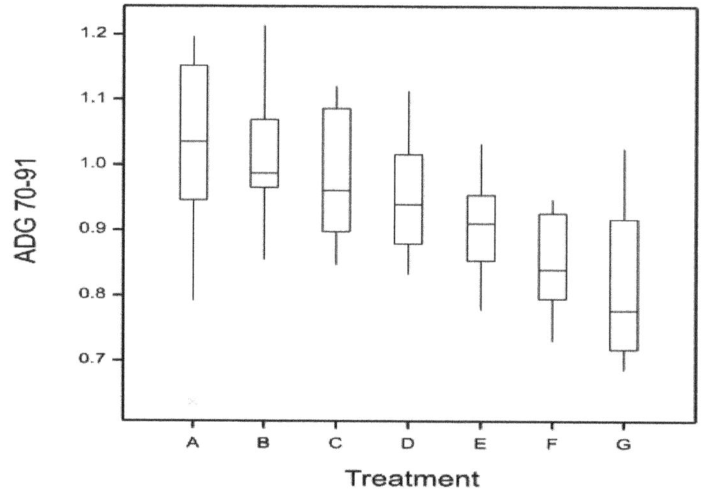

Figure 4. Box and whisker plot of ADG between day 70 and 91 for treatments A to G (Enterisol vaccine plus lime; vaccine only; lime only; no treatment; 12 ppm; 25 ppm; and 50 ppm olaquindox). Descriptive statistics were not tested for significance. Green cross indicates outlier.

In this final period, there was a strong negative correlation between *L. intracellularis* numbers at day 84 and the ADG ($R = -0.573$ and $p < 0.001$) and ADFI ($R = -0.616$ and $p < 0.001$) and a positive correlation with the FCR ($R = 0.276$ and $p = 0.011$).

3.6. Epidemiology of PE in Pigs from 9 to 21 Weeks of Age (Trial Days 0 to 84)

Over the whole trial period, the ADG was significantly different between treatments ($p < 0.001$) (Table 4) and between gender (males = 0.8671 and females = 0.8327, $p < 0.001$)

(Figure 5). The starting weight, replicate, and interactions between treatment and gender did not significantly affect the ADG. The ADFI was significantly different between treatments ($p = 0.003$), with the highest feed intake in the vaccinated plus lime treatment A and the lowest ADFI in the 25 ppm olaquindox treatment (Table 4). The replicate, gender, and starting weight also had a significant effect on the ADFI, with a higher intake in males relative to females (1.945 and 1.912, respectively). The FCR was not significantly different between treatments over the whole period ($p = 0.154$), but the replicate, gender, starting weight, and treatment–gender interactions did significantly impact the FCR ($p < 0.001$). Males showed a lower FCR than females (2.204 versus 2.337).

Table 4. Predicted ADG, ADFI, and feed conversion efficiency (between day 0 and 91) blocked by replicate and gender with starting weight as a covariate (ANOVA), with standard error of difference (Av SED).

Treatment Group	Treatment Description	ADG	ADFI	FCR
A	Disinfectant + vaccination	0.8714 ^{a}	1.990 ^{a}	2.287
B	Vaccination	0.8640 ^{ab}	1.941 ^{ac}	2.248
C	Disinfectant	0.8574 ^{ab}	1.934 ^{bc}	2.256
D	No treatment	0.8453 ^{bc}	1.926 ^{bc}	2.280
E	12 ppm Olaquindox	0.8436 ^{bc}	1.938 ^{abc}	2.298
F	25 ppm Olaquindox	0.8280 ^{c}	1.888 ^{b}	2.281
G	50 ppm Olaquindox	0.8312 ^{c}	1.894 ^{bc}	2.280
Av SED		0.01033	0.02602	0.02216

Different superscripts within columns indicate significant differences.

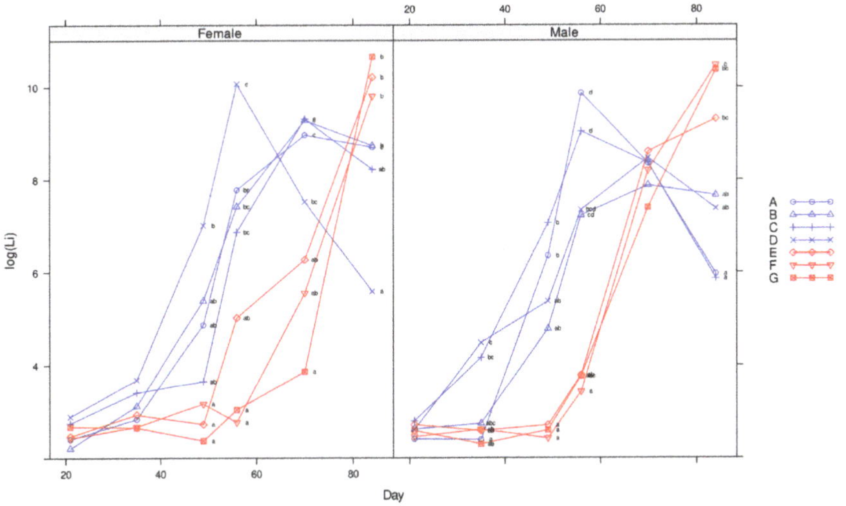

Figure 5. Mean Log_{10} *L. intracellularis* numbers excreted over time for each gender (females on left hand side and males on right side) and treatments A to G (Enterisol vaccine plus lime; vaccine only; lime only; no treatment; 12 ppm; 25 ppm; and 50 ppm olaquindox). Different superscripts indicate significant differences.

The pig liveweight at 91 days and the mean P2 backfat was affected by treatment ($p = 0.012$ and 0.015, respectively) with lower P2 values and lower liveweights in pigs medicated with higher concentrations of olaquindox (Table 5). Carcass measures, including hot standard carcass weight (HSCW) and mean dressing percent, were not affected by treatment ($p > 0.05$). The percentage of intestines condemned at abattoir due to damage from *L. intracellularis* infection was between 0% and 4.48% but was only recorded for the 783 pigs slaughtered in replicates 5 and 6, so the results were not analyzed for significance.

The carcass dressing percent, P2 backfat, HSCW, and mean liveweight at 91 days were all affected by the replicate ($p < 0.012$), and all measures except HSCW were also affected by gender ($p < 0.044$).

Table 5. Predicted mean liveweight, mean backfat, and percent condemned intestines at slaughter (day 91) blocked by replicate and gender with starting weight as a covariate (ANOVA), with average standard error of difference (Av SED).

Treatment Group	Treatment Description	Mean Liveweight 91 Days (kg)	Mean P2 Backfat (mm)	% Condemned Intestines *
A	Disinfectant + vaccination	102.3 ^{a}	12.34 ^{a}	2.29
B	Vaccination	101.5 ^{ab}	12.14 ^{ad}	0.85
C	Disinfectant	100.6 ^{abd}	12.41 ^{a}	1.56
D	No treatment	99.9 ^{bc}	11.71 ^{bd}	0
E	12 ppm Olaquindox	100.0 ^{bc}	11.88 ^{abd}	1.96
F	25 ppm Olaquindox	98.7 ^{cd}	11.49 ^{bc}	2.53
G	50 ppm Olaquindox	98.0 ^{c}	11.89 ^{abd}	4.48
Av SED		1.165	0.3066	

Different superscripts within columns indicate significant differences. * % intestines condemned due to intestinal hemorrhage only scored for replicates 5 and 6.

4. Discussion

Molecular and serological epidemiology tools were able to demonstrate when pens of commercial pigs became infected with *L. intracellularis,* the severity of PE lesions by inference from the numbers of excreted *L. intracellularis*, and the proportion and timing of pigs that raised an immune response to infection. Clinical signs correlated in time with peaks in *L. intracellularis* excretion as previously demonstrated [31], and significant negative correlations were observed between the ADG and excretion of *L. intracellularis* in the periods 21 to 56 days, 56 to 70 days, and 70 to 84 days. Similar negative correlations between the ADG and number of excreted *L. intracellularis* have been reported in fecal samples from individual pigs [33], but this is the first report of this correlation for pens of pigs on commercial farms. Ultimately, these epidemiological tools enabled the comparison of treatment options for the control of PE in a large commercial pig herd.

Medicating all weaner pigs with olaquindox from 3 to 9 weeks of age prevented *L. intracellularis* infection in all pigs prior to the start of the trial, as demonstrated by the absence of *L. intracellularis*-specific antibodies 30 days after treatments commenced. Serum IgG antibodies are routinely detected 14 to 28 days post exposure, depending on the oral dose of *L. intracellularis* [4,41]. As a consequence, this study evaluated PE control treatments in pens of commercially reared pigs between 9 and 22 weeks of age. Pigs in this study were naturally exposed to a relatively low number of *L. intracellularis* after 9 weeks of age, as demonstrated by the low level of *L. intracellularis* excretion in untreated control pigs at 14 weeks of age and the absence of significant production losses in this group between 9 and 12 weeks of age. Clinical PE and the excretion of higher numbers of *L. intracellularis* were not demonstrated in control pigs until they reached 15 to 16 weeks of age (6 to 7 weeks after the trial commenced). In experimental *L. intracellularis* challenge trials, clinical signs of PE and production losses are usually evident between 3 and 4 weeks after exposure [4,42]. In this commercial piggery study, medication, vaccination, and disinfection treatments all impacted on the timing and severity of *L. intracellularis* infection and the associated clinical signs of PE.

Most pigs maintained on high levels of in-feed olaquindox remained naïve to *L. intracellularis* until the medication was removed from their diets, as evidenced by the absence of detectable *L. intracellularis* antibodies at day 63 and low fecal excretion of *L. intracellularis* at days 35, 49, and 56. This agrees with experimental challenge studies, where most pigs medicated with 25 or 50 ppm olaquindox failed to excrete *L. intracellularis* in their feces and were, therefore, susceptible to *L. intracellularis* infection once olaquindox was removed from their diets [29]. Once medication was removed from our commercial herd, naïve finisher pigs developed severe clinical signs, including black tarry feces and death due to intestinal hemorrhage, and they excreted extremely high numbers of *L. intracellularis* (10^{10} per gram of feces compared with the next highest of $10^{8.8}$). Net revenue is significantly impacted by the death of finisher pigs, where the cumulative costs of feeding pigs from weaning to finishing cannot be recovered. In addition, pigs affected with PHE at 19 to 21 weeks of age did not have sufficient time to recover weight losses before slaughter, and up to 4.5% of their intestines could not be sold for sausage casing. Outbreaks of severe PHE are reported to be associated with the increasing age of infected pigs, rather than any difference in the challenge dose of *L. intracellularis* [6,7]. A range of other antibiotics at high concentrations have been reported to prevent *L. intracellularis* infection, including tiamulin [24], chlortetracycline and oxytetracycline [43], tylosin phosphate [25], and tilmicosin [44]. Field and experimental challenge studies have demonstrated that *L. intracellularis* infection of naïve finisher pigs after the removal of protective antibiotics can cause PHE outbreaks [7,45].

Interestingly, pigs medicated with 12 ppm olaquindox started to excrete higher numbers of *L. intracellularis* at an earlier time than pigs medicated with higher concentrations of olaquindox, suggesting that 12 ppm was not as effective in preventing *L. intracellularis* infection compared with higher levels. As a consequence, pigs medicated with 12 ppm olaquindox were protected from severe clinical signs of hemorrhagic enteropathy and production losses. The ADG between 9 and 22 weeks of age in the pigs medicated with 12 ppm olaquindox was not significantly different to control pigs, vaccinated pigs, and pigs housed in lime-treated pens.

Neither vaccination alone nor disinfection alone significantly increased the ADG or ADFI nor decreased the feed to gain relative to control pigs over the grower and finisher phases of production, contrary to previous vaccination studies in experimental challenge trials [20]. However, housing vaccinated pigs in lime-treated pens did significantly improve production measures relative to control pigs. The vaccination of pigs in washed or washed and disinfected pens appeared to delay or suppress *L. intracellularis* infection, observed as significantly fewer *L. intracellularis* excreted at day 35 compared with control pigs. However, by days 49 and 56, there were no significant differences in the number of *L. intracellularis* excreted, and by day 70, vaccinated pigs shed significantly more *L. intracellularis* than control pigs. In contrast, vaccination was able to reduce the fecal excretion of *L. intracellularis* in experimental studies where pigs were challenged with *L. intracellularis* at a single time point three weeks after vaccination [20,41]. Under commercial production conditions, pigs may be challenged with *L. intracellularis* many times, and pigs in this trial were vaccinated at least 6 weeks before they were exposed to *L. intracellularis*. Both vaccination and lime disinfection only appeared to suppress *L. intracellularis* infection for a short period around the time of exposure. Challenging pigs with a lower dose of *L. intracellularis* can also suppress *L. intracellularis* infection and clinical disease [3,5,7]. This study suggests that once *L. intracellularis* enter the pigs' intestinal epithelium and start replicating, antibiotics that can accumulate intracellularly are best at suppressing infection and disease [45].

L. intracellularis is transmitted between pigs by the fecal-oral route [16,17], so poor pen hygiene and the accumulation of manure increase the dose of *L. intracellularis* that pigs are exposed to. In addition, *L. intracellularis* can survive in feces for at least two weeks and colonize naïve weaner pigs dosed with contaminated feces [16,35]. On-farm epidemiology studies demonstrated that sheds with poor hygiene due to high stocking densities, a continuous flow production, and partially slatted floors increased the risk of *L. intracellularis* infection, while all-in-all production and the use of disinfectants between batches of pigs appeared to protect pigs [13,18,19,46]. Disinfectants like Virkon S can successfully eliminate *L. intracellularis* from contaminated grower pens and can prevent transmission of *L. intracellularis* to naïve pigs in disinfected pens [35]. However, lime disinfection of commercial grower pens failed to significantly reduce *L. intracellularis* excretion over the following 13 weeks relative to other treatments in this study. Housing pigs in lime-disinfected pens also failed to improve pig production measures associated with PE relative to control pigs, which were repeatedly exposed to *L. intracellularis*.

In contrast, the disinfection of pens in combination with vaccination increased the protection of pigs against PE. Combining the disinfection of commercial pens with antibiotic medication and the partial depopulation of herds has also proved successful in reducing the survival and transmission of *L. intracellularis* in attempts to eradicate PE from small commercial herds [47–49].

Under commercial conditions, where pigs were housed in washed but not disinfected pens throughout the grower and finisher periods, antibiotic pulses were also required to control clinical signs of PE. Once diarrhea was observed, pigs in all treatments were medicated in feed with tylosin two days of every 10 days from day 56 to 84. Previous studies demonstrate that tylosin can suppress clinical signs of PE and reduce the fecal excretion of *L. intracellularis* and the associated production losses [25,29].

For molecular and serological monitoring of commercial herds to be economical, some thought needs to be given to the optimal timing for sample collection. The target sampling period should only commence when preventative antibiotics are removed from diets and before clinical signs of PE are evident. Sampling should continue to the end of the finisher phase. As serological monitoring aims to demonstrate both the timing of the *L. intracellularis* challenge and the proportion of pigs with immunity, blood needs to be collected from about 10% of pigs every three to four weeks. Published *L. intracellularis* serological surveys support this frequency of sampling to identify when pigs are exposed to *L. intracellularis* and to determine the seroprevalence of herds [10]. Pooled pen fecal samples probably need to be collected every two to three weeks, as experimental challenge studies indicate that the fecal excretion of *L. intracellularis* persists for about four weeks, depending on the oral challenge dose [3,4]. In this study, pooled pen fecal samples were collected from every pen every two weeks because the impact of seven different treatments was being investigated over the same period. It may have been possible to reduce the number of sampled pens per treatment in those treatments with 14 pens. This sampling protocol provided sufficient statistical power to demonstrate the efficacy of PE control measures on a commercial herd with 8 to 14 pens per treatments, given the experimental design outlined.

Evaluating PE control on commercial farms can be confounded by the presence of other diseases in the herd. In this study, outbreaks of pleuropneumonia were observed in pigs between 9 and 15 weeks of age, and swine dysentery and PE outbreaks were observed between 15 and 22 weeks of age. As olaquindox is also effective in controlling swine dysentery [50], some of the production gains in olaquindox-treated pigs between 9 and 17 weeks of age may have been due to better control of swine dysentery as well as PE. However, studies have also reported the recurrence of swine dysentery once olaquindox is removed [50], as observed in treatments E, F, and G.

5. Conclusions

Molecular and serological epidemiology tools have been successfully used to identify when pigs in commercial herds are exposed to *L. intracellularis*, the severity of PE they experience, when the majority of pigs have developed immunity, and when treatments are required. Vaccination and the cleaning and disinfection of pens before pigs enter grower facilities can reduce exposure to *L. intracellularis* and can reduce clinical signs of PE and the associated production losses, but PE control may also require antibiotic medication if clinical signs persist. Molecular and serological epidemiology tools can identify the optimal timing for a wide range of treatments. As the pig industry aims to identify management practices that reduce its reliance on antibiotics to control bacterial diseases, this trial has demonstrated a method to evaluate PE control on the farm and also demonstrated that PE can be controlled by a combination of good hygiene and vaccination, with the judicious use of antibiotics if required.

Author Contributions: A.C. and C.C. contributed equally to conceptualization, methodology, validation, formal analysis, investigation and data curation. A.C. contributed primarily to the preparation of the original draft, with A.C. and C.C. reviewing and editing the drafts. Supervision, A.C. and C.C.; Project Administration, A.C. and C.C.; Funding Acquisition, A.C. and C.C. All authors have read and agreed to the published version of the manuscript.

Funding: This research was supported financially by the Cooperative Research Centre for High Integrity Australian Pork, grant number 2C-102 and the New South Wales Department of Primary Industries.

Institutional Review Board Statement: The study was conducted according to the Australian Code for the Care and Use of Animals for Scientific Purposes, and animal ethics approval was granted by the Animal Ethics Committees of the commercial pig production company and Elizabeth Macarthur Agricultural Institute (13V057C).

Informed Consent Statement: The owners of the pigs supplied provided informed consent for the use of these animals through an animal ethic committee.

Data Availability Statement: Data is unavailable due to privacy restrictions.

Acknowledgments: The authors wish to acknowledge the veterinary support of Peter McKenzie, the biometry support provided by Damian Collins and the technical support provided by Steven Heavener, Leanne Bringolf, Shayne Fell (NSW DPI) and Kirrily O'Halloran (piggery).

Conflicts of Interest: Author Cheri Collins was employed by the company Rivalea Australia. The remaining authors declare that the research was conducted in the absence of any commercial or financial relationships that could be construed as a potential conflict of interest. The funders had no role in the design of the study; in the collection, analyses, or interpretation of data; in the writing of the manuscript; or in the decision to publish the results.

References

1. McOrist, S.; Jasni, S.; Mackie, R.A.; Macintyre, N.; Neef, N.; Lawson, G.H.K. Reproduction of porcine proliferative enteropathy with pure cultures of Ileal symbiont intracellularis. *Infect. Immun.* **1993**, *61*, 4286–4292. [CrossRef] [PubMed]
2. Lawson, G.H.K.; Gebhart, C.J. Proliferative enteropathy. *J. Comp. Pathol.* **2000**, *122*, 77–100. [CrossRef]
3. Guedes, R.; Winkelman, N.; Gebhart, C. Relationship between the severity of porcine proliferative enteropathy and the infectious dose *Lawsonia intracellularis*. *Vet. Rec.* **2003**, *153*, 432–433. [CrossRef] [PubMed]
4. Collins, A.M.; Love, R.J. Re-challenge of pigs following recovery from proliferative enteropathy. *Vet. Microbiol.* **2007**, *120*, 381–386. [CrossRef] [PubMed]
5. Paradis, M.A.; Gebhart, C.J.; Toole, D.; Vessie, G.; Winkelman, N.L.; Bauer, S.A.; Wilson, J.B.; McClure, C.A. Sub-clinical ileitis: Diagnostic and performance parameters in a multi-dose mucosal homogenate challenge model. *J. Swine Health Prod.* **2012**, *20*, 137–141.
6. Love, R.J.; Love, D.N.; Edwards, M.J. Proliferative haemorrhagic enteropathy in pigs. *Vet. Rec.* **1977**, *100*, 65–68. [CrossRef]
7. Collins, A.M.; McOrist, S.; van Dijk, M.; Love, R.J. Effect of age on clinical disease associated with *Lawsonia intracellularis* infection. In *Manipulating Pig Production;* Cranwell, P.D., Ed.; Australasian Pig Science Association: Werribee, Australia, 1999; p. 277.
8. Holyoake, P.K.; Collins, A.; Mullan, B. Simulation of the Economic Impact of *Lawsonia intracellularis* Infection. In Proceedings of the 21st International Pig Veterinary Society Congress, Vancouver, BC, Canada, 18–21 July 2010; p. 233.
9. Corzo, C.A.; Friendship, R.M.; Dewey, C.E.; Blackwell, T. Seroprevalence of *Lawsonia intracellularis* in Ontario swine herds. *J. Swine Health Prod.* **2005**, *13*, 314–317.
10. Holyoake, P.K.; Emery, D.; Gonsalves, J.; Donahoo, M.; Collins, A. Prevalence of antibodies to *Lawsonia intracellularis* in pig herds in Australia. *Aust. Vet. J.* **2010**, *88*, 186–188. [CrossRef]
11. Pozo, J.; Collins, A.M.; Rubio, P.; Love, R.J. Maternal Immunity to *Lawsonia intracellularis* Infection. In Proceedings of the 16th International Pig Veterinary Society Congress, Melbourne, Australia, 17–21 September 2000; Cargill, C., McOrist, S., Eds.; Causal Productions Pty Limited: Melbourne, Australia, 2000; p. 108.
12. Bane, D.; Norby, B.; Gardner, I.; Roof, M.; Knittel, J.P.; Bush, E.J. Prevalence of Management Risk Factors Associated with *Lawsonia intracellularis* Seropositivity in the US Swine Herd. In Proceedings of the Allen D Leman Swine Conference, University of Minnesota, Minneapolis, MN, USA, 21 September 1997; pp. 10–11.
13. Bronsvoort, M.; Norby, B.; Bane, D.P.; Gardner, I.A. Management factors associated with seropositivity to *Lawsonia intracellularis* in US swine herds. *J. Swine Health Prod.* **2001**, *9*, 285–290.
14. Paradis, M.A.; Gottschalk, M.; Rajic, A.; Ravel, A.; Wilson, J.B.; Aramini, J.; McClure, C.A.; Dick, C.P. Seroprevalence of *Lawsonia intracellularis* in different swine populations in 3 provinces in Canada. *Can. Vet. J.* **2007**, *48*, 57–62.
15. Bae, J.K.; Wieland, B.; Sait, M.; Longbottom, D.; Smith, D.G.E.; Alarcon, P.; Wheelhouse, N. Risk factors associated with *Lawsonia intracellularis* in English pig farms. *Vet. J.* **2013**, *197*, 707–711. [CrossRef] [PubMed]
16. Collins, A.; Love, R.J.; Pozo, J.; Smith, S.H.; McOrist, S. Studies on the ex vivo survival of *Lawsonia intracellularis*. *Swine Health Prod.* **2000**, *8*, 211–215.

17. Jordan, D.M.; Knittel, J.P.; Schwartz, K.J.; Roof, M.; Hoffman, L.J. A *Lawsonia intracellularis* transmission study using a pure culture inoculated seeder-pig sentinel model. *Vet. Microbiol.* **2004**, *104*, 83–90. [CrossRef] [PubMed]
18. Bane, D.P.; Neumann, E.; Gebhart, C.J.; Gardner, I.A.; Norby, B. Porcine proliferative enteropathy: A case-control study in swine herds in the United States. *J. Swine Health Prod.* **2001**, *9*, 155–158.
19. Collins, A.M.; Love, R.J. Risk Factors Associated with *Lawsonia intracellularis* Infection. In *Manipulating Pig Production*; Paterson, J.E., Ed.; Australasian Pig Science Association: Werribee, Australia, 2003; p. 34.
20. Kroll, J.J.; Roof, M.; McOrist, S. Evaluation of protective immunity in pigs following oral administration of an avirulent live vaccine of *Lawsonia intracellularis*. *Am. J. Vet. Res.* **2004**, *65*, 559–565. [CrossRef] [PubMed]
21. Guedes, R.M.C.; Gebhart, C.J. Evidence of cell-mediated immune response and specific local mucosal immunoglobulin (Ig) A production against *Lawsonia intracellularis* in experimentally infected swine. *Can. J. Vet. Res.* **2010**, *74*, 97–101.
22. Cordes, H.; Riber, U.; Jensen, T.K.; Jungersen, G. Cell-mediated and humoral immune responses in pigs following primary and challenge-exposure to *Lawsonia intracellularis*. *Vet. Res.* **2012**, *43*, 9. [CrossRef] [PubMed]
23. Nogueira, M.G.; Collins, A.M.; Donahoo, M.; Emery, D. Immunological responses to vaccination following experimental *Lawsonia intracellularis* virulent challenge in pigs. *Vet. Microbiol.* **2013**, *164*, 131–138. [CrossRef] [PubMed]
24. McOrist, S.; Smith, S.H.; Shearn, M.F.H.; Carr, M.M.; Miller, D.J.S. Treatment and prevention of porcine proliferative enteropathy with oral Tiamulin. *Vet. Rec.* **1996**, *139*, 615–618.
25. McOrist, S.; Morgan, J.; Veenhuizen, M.F.; Lawrence, K.; Kroger, H.W. Oral administration of Tylosin phosphate for treatment and prevention of proliferative enteropathy in pigs. *Am. J. Vet. Res.* **1997**, *58*, 136–139. [CrossRef]
26. Winkelman, N.L.; Gebhart, C.; Collins, J.; King, V.; Hannon, M.; Wolff, T. An Evaluation of Aureomycin Chlortetracycline (CTC) Granular Feed Additive for Prevention or Treatment of Swine Ileitis. In Proceedings of the 28th Annual Meeting of the American Association of Swine Practitioners, Quebec City, QC, Canada, 1–4 March 1997; pp. 79–83.
27. Winkelman, N.L.; Cornell, P.; Bradford, J. Lincomycin Feed Medication and Two Water Medications Against Ileitis Caused by *Lawsonia intracellularis*. In Proceedings of the 29th Annual Meeting of the American Association of Swine Practitioners, Des Moines, IA, USA, 7–10 March 1998; pp. 195–199.
28. Schwartz, K.J.; Knittel, J.P.; Walter, D.; Roof, M.; Anderson, M. Effect of oral tiamulin on the development of porcine proliferative enteropathy in a pure-culture challenge model. *Swine Health Prod.* **1999**, *7*, 5–11.
29. Vu, N.Q.; Collins, A.M.; van Dijk, M.; Love, R. Effect of Antibiotic Medication on the Development of Immunity to *Lawsonia intracellularis* in Pigs. In *Manipulating Pig Production*; Cranwell, P.D., Ed.; Australasian Pig Science Association: Werribee, Australia, 2001; p. 148.
30. Richter, B.; Ladinig, A.; Nedorost, N.; Weissenbock, H. A TaqMan quantitative polymerase chain reaction assay for the detection of *Lawsonia intracellularis* in fecal and tissue samples from pigs. *J. Vet. Diagn. Investig.* **2010**, *22*, 70–73. [CrossRef] [PubMed]
31. Pedersen, K.S.; Skrubel, R.; Stege, H.; Angen, O.; Stahl, M.; Hjulsager, C.; Larsen, L.E.; Nielsen, J.P. Association between average daily gain, faecal dry matter content and concentration of *Lawsonia intracellularis* in faeces. *Acta Vet. Scand.* **2012**, *54*, 58. [CrossRef] [PubMed]
32. Johansen, M.; Nielsen, M.; Dahl, J.; Svensmark, B.; Baekbo, P.; Kristensen, C.S.; Hjulsager, C.K.; Jensen, T.K.; Stahl, M.; Larsen, L.E.; et al. Investigation of the association of growth rate in grower-finishing pigs with the quantification of *Lawsonia intracellularis* and porcine circovirus type 2. *Prev. Vet. Med.* **2013**, *108*, 63–72. [CrossRef] [PubMed]
33. Collins, A.M.; Barchia, I.M. The critical threshold of *Lawsonia intracellularis* in pig faeces that causes reduced average daily gains in experimentally challenged pigs. *Vet. Microbiol.* **2014**, *168*, 455–458. [CrossRef] [PubMed]
34. Love, R.J. Gastro Intestinal Disorders of Growers. In *Gastroenteric Conditions Seminar*; Post-Graduate Committee in Veterinary Science, Ed.; The University of Sydney: Sydney, Australia, 1981; pp. 397–401.
35. Collins, A.M.; Fell, S.A.; Barchia, I.M. Cleaning and disinfection with Virkon S significantly reduces *Lawsonia intracellularis* survival and transmission to naive pigs. *J. Swine Health Prod.* **2013**, *21*, 144–147.
36. Collins, A.M.; Barchia, I.M. Optimal Pooling of Faeces to Quantify *Lawsonia intracellularis* in Clinically and Sub-Clinically Affected Pigs. In *Manipulating Pig Production XIV*; Pluske, J.A.J., Ed.; Australasian Pig Science Association: Werribee, Australia, 2013; p. 200.
37. Collins, A.M.; Fell, S.; Pearson, H.; Toribio, J.A. Colonisation and shedding of *Lawsonia intracellularis* in experimentally inoculated rodents and in wild rodents on pig farms. *Vet. Microbiol.* **2011**, *150*, 384–388. [CrossRef] [PubMed]
38. Brooks, M.E.; Kristensen, K.; van Benthem, K.J.; Magnusson, A.; Berg, C.W.; Nielsen, A.; Skaug, H.J.; Maechler, M.; Bolker, B.M. glmmTMB balances speed and flexibility among packages for zero-inflated generalized linear mixed modeling. *R J.* **2017**, *9*, 378–400. [CrossRef]
39. Hothorn, T.; Bretz, F.; Westfall, P. Simultaneous inference in general parametric models. *Biom. J.* **2008**, *50*, 346–363. [CrossRef]
40. Lenth, R. Emmeans: Estimated Marginal Means, Aka Least-Squares Means. R Package Version 1.8.0. 2022. Available online: https://CRAN.R-project.org/package=emmeans (accessed on 24 March 2024).
41. Guedes, R.M.C.; Gebhart, C.J. Onset and duration of fecal shedding, cell-mediated and humoral immune responses in pigs after challenge with a pathogenic isolate or attenuated vaccine strain of *Lawsonia intracellularis*. *Vet. Microbiol.* **2003**, *91*, 135–145. [CrossRef]
42. Guedes, R.M.C.; Gebhart, C.J. Comparison of intestinal mucosa homogenate and pure culture of the homologous *Lawsonia intracellularis* isolate in reproducing proliferative enteropthy in swine. *Vet. Microbiol.* **2003**, *93*, 159–166. [CrossRef] [PubMed]

43. McOrist, S.; Shearn, M.F.; Morgan, J. Control of porcine proliferative enteropathy by oral administration of chlortetracycline. *Vet. Rec.* **1999**, *144*, 48–49. [CrossRef] [PubMed]
44. McOrist, S.; Mackie, R.A.; Lawson, G.H. Antimicrobial susceptibility of ileal symbiont intracellularis isolated from pigs with proliferative enteropathy. *J Clin Microbiol.* **1995**, *33*, 1314–1317. [CrossRef] [PubMed]
45. Beers, P. Studies on Porcine Adenomatosis with Particular Reference to Proliferative Haemorrhagic Enteropathy. Ph.D. Dissertation, Univeristy of Sydney, Sydney, Australia, 1984.
46. Pearce, G.P. Epidemiology of enteric disease in grower and finisher pigs: A postal survey of pig producers in England. *Vet. Rec.* **1997**, *144*, 338–342. [CrossRef] [PubMed]
47. Flø, H.; Bock, R.; Oppengaard, O.J.; Bergsjø, B.; Lium, B. An Attempt to Eradicate *Lawsonia intracellularis* and *Brachyspira* sp. from Swine Herds. In Proceedings of the 16th International Pig Veterinary Society Congress, Melbourne, Australia, 17–21 September 2000; p. 66.
48. Johansen, M.; Baekbo, P.; Jensen, T.K.; Møller, K.; Nielsen, V.R. Attempt to Eradicate *Lawsonia intracellularis* by Medication in 9 Sow Herds—Preliminary Results. In Proceedings of the 17th International Pig Veterinary Society Congress, Ames, IA, USA, 2–5 June 2002.
49. Ellegaard, B.; Koerting, K.; Szancer, J. Attempt to Eradicate *Lawsonia intracellularis* from an SPF Farm by Medication. In Proceedings of the 20th International Pig Veterinary Society Congress, Durban, South Africa, 22–25 June 2008; p. 152.
50. Raynaud, J.P.; Brunault, G.; Patterson, E.B. A swine dysentery model for evaluation of drug prophylaxis: Efficacy of various drugs in the control of swine dysentery. *Am. J. Vet. Res.* **1981**, *42*, 51–53. [PubMed]

animals

MDPI

Article

Surveillance of Sarcoptic Mange in Iberian Ibexes (*Capra pyrenaica*) and Domestic Goats (*Capra hircus*) in Southern Spain

Félix Gómez-Guillamón [1,†], Débora Jiménez-Martín [2,†], Debora Dellamaria [3], Antonio Arenas [2], Luca Rossi [4], Carlo V. Citterio [3], Leonor Camacho-Sillero [1], Barbara Moroni [5], David Cano-Terriza [2,6,*] and Ignacio García-Bocanegra [2,6]

1 Programa de Vigilancia Epidemiológica de la Fauna Silvestre (PVE), Consejería de Sostenibilidad, Medio Ambiente y Economía Azul, Junta de Andalucía, 29002 Málaga, Spain; felixj.gomezguillamon@juntadeandalucia.es (F.G.-G.); leonorn.camacho@juntadeandalucia.es (L.C.-S.)

2 Departamento de Sanidad Animal, Grupo de Investigación en Sanidad Animal y Zoonosis (GISAZ), UIC Zoonosis y Enfermedades Emergentes ENZOEM, Universidad de Córdoba, 14014 Córdoba, Spain; debora.djm@gmail.com (D.J.-M.); arenas@uco.es (A.A.); nacho.garcia@uco.es (I.G.-B.)

3 Istituto Zooprofilattico Sperimentale delle Venezie, 35020 Legnaro, Italy; ddellamaria@izsvenezie.it (D.D.); ccitterio@izsvenezie.it (C.V.C.)

4 Department of Veterinary Sciences, University of Turin, 10095 Grugliasco, Italy; luca.rossi@unito.it

5 Istituto Zooprofilattico Sperimentale del Piemonte, Liguria e Valle d'Aosta, via Bologna 148, 10154 Torino, Italy; barbara.moroni@izsto.it

6 CIBERINFEC, ISCIII—CIBER de Enfermedades Infecciosas, Instituto de Salud Carlos III, 28029 Madrid, Spain

* Correspondence: v82cated@uco.es

† These authors contributed equally to the work.

Simple Summary: A serosurvey study was conducted in southern Spain to assess the exposure and spatial distribution of *Sarcoptes scabiei* in Iberian ibexes (*Capra pyrenaica*) and domestic goats (*Capra hircus*). The study included sera from 411 Iberian ibexes (157 with skin lesions compatible with sarcoptic mange and 254 that were clinically healthy), skin samples from 88 affected animals, and 392 serum samples from domestic goats, collected between 2015 and 2021. Antibodies against *S. scabiei* were found in 3.1% of the clinically healthy ibexes and 66.2% of those with compatible skin lesions. Mites were confirmed in 64.8% of the skin samples, and 86.0% of these mite-positive individuals had antibodies. Seropositive animals were detected in population nuclei with previous records of sarcoptic mange, but not in historically free population nuclei. The non-detection of antibodies against *S. scabiei* in the domestic goats suggests an independent epidemiological cycle of sarcoptic mange in Iberian ibex populations in the study area. Integrated surveillance programs and control strategies in wildlife and livestock are essential to mitigating the risk of *S. scabiei* circulation in Iberian ibex populations.

Abstract: Sarcoptic mange is a highly contagious skin disease caused by *Sarcoptes scabiei*. Sera were collected from 411 Iberian ibexes, comprising 157 individuals with sarcoptic mange skin lesions and 254 clinically healthy animals, in 13 population nuclei across Andalusia (southern Spain) between 2015 and 2021. Skin samples from 88 of the 157 animals with mange-compatible lesions were also obtained. Moreover, 392 serum samples from domestic goats (*Capra hircus*) were collected in the same region and study period. Antibodies against *S. scabiei* were tested using an in-house indirect ELISA, while the presence of mites of *S. scabiei* was evaluated in the skin samples by potassium hydroxide digestion. Seropositivity was found in eight (3.1%) of the clinically healthy ibexes and in 104 (66.2%) of the animals with mange-compatible lesions. The presence of *S. scabiei* was confirmed in 57 (64.8%) out of the 88 skin samples analysed and anti-*S. scabiei* antibodies were found in 49 (86.0%) of these 57 mite-positive individuals. Seropositive animals were detected in population nuclei with previous records of sarcoptic mange, where *S. scabiei* mites were detected by potassium hydroxide digestion in individuals with sarcoptic mange-compatible external lesions. However, seropositivity was not observed in population nuclei that were historically free of this disease. None of the 392 domestic goats had antibodies against *S. scabiei*, suggesting an independent epidemiological cycle of sarcoptic mange in Iberian ibex populations in the study area, and a limited or null role of domestic goats in the transmission of the parasite to this wild species. Overall, our findings underscore the importance of maintaining and/or implementing integrated surveillance programs and control strategies in wildlife and livestock, to limit the risk of *S. scabiei* circulation in Iberian ibex populations.

Keywords: Caprinae; domestic goat; mite; *Sarcoptes*; wild goat; monitoring

1. Introduction

Sarcoptic mange, caused by the obligate burrowing mite *Sarcoptes scabiei*, is a highly contagious skin disease distributed worldwide [1,2]. This mite is considered one of the terrestrial ectoparasites with the widest host range, affecting both wildlife and domestic animals, and even humans [3]. In fact, *Sarcoptes scabiei* currently affects more than 100 mammalian species worldwide, posing an ongoing challenge to wildlife conservation and management in particular, as it has been associated with significant declines in local populations for decades [4].

Clinical manifestations of sarcoptic mange include intense pruritus, hair loss, scaling, and crusting or hyperkeratosis, among others [5]. Although the severity and clinical outcomes of sarcoptic mange can differ among species, populations, and individuals [6], it remains likely to be the most severe disease affecting wild Caprinae in Europe [7,8].

The Iberian ibex, or Iberian wild goat (*Capra pyrenaica*), is a distinctive endemic medium-sized wild ruminant emblematic of the Iberian Peninsula. Traditionally, four different subspecies of Iberian ibex have been described: *C. pyrenaica lusitanica*, which formerly inhabited northern Portugal and certain regions of northwestern Spain; *C. p. pyrenaica*, in the Pyrenees Mountains; *C. p. hispanica*, found in the southern and eastern regions of the Iberian Peninsula; and *C. p. victoriae*, distributed mainly in central areas of Spain [9]. At present, only two of the four subspecies originally described are not extinct (*C. p. hispanica* and *C. p. victoriae*) with a population of around 100,000 individuals that is expanding, or at least stable, in southern, central, and eastern regions of Spain, with smaller, localized populations also found in northern Portugal and southern France [10].

The Iberian ibex has been shown to be particularly sensitive to *S. scabiei* infection, which results in high rates of morbidity and mortality, especially in naïve populations [7]. In this regard, the first known epizootic outbreak of sarcoptic mange that affected this species caused an over 95% decline in an estimated population of almost 9500 individuals [11,12]. The introduction of infected domestic goats was the most likely origin of this outbreak [4, 12]. During the decade following this first outbreak, the disease spread throughout the mountain range and into surrounding areas [6]. This epidemiological transition from an initial epizootic outbreak to an endemic disease has been observed in most of the affected Iberian ibex populations [12–15]. However, the prevalence and mortality rates associated with sarcoptic mange vary between geographic locations and populations [16–18].

Given that sarcoptic mange is a concern for both animal health and conservation in the Iberian ibex, several studies have been conducted in order to better understand different aspects of the disease in this species, including ecology, physiology, pathology, genetics, control strategies, and diagnostic methods, among others [14]. However, no large-scale epidemiological studies have been carried out to evaluate the circulation of this parasite in Iberian ibex and sympatric domestic goat populations. The aim of the present study is to assess the exposure and spatial distribution of *S. scabiei* in Iberian ibex and domestic goat populations in Andalusia (southern Spain), the Spanish region with the largest census of both species.

2. Materials and Methods

2.1. Study Area and Data Collection

During 2015–2021, 411 blood samples from culled Iberian ibexes were collected in the framework of the epidemiological surveillance program coordinated by the Regional Government of Andalusia [19]. Andalusia (southern Spain: 36° N–38°60′ N, 1°75′ W–7°25′ W), the study region, has a predominantly warm Mediterranean climate: mild winters with irregular precipitation, and dry, hot, sunny summers along the coast, becoming more extreme further inland. The average annual temperature is around 18 °C, with over 300 days of sunshine per year. January is the coldest month, while August records the highest temperatures.

Iberian ibexes were opportunistically sampled from the primary population nuclei of this species (n = 13), specifically, those with more than one individual per square kilometre. Ten of them were population nuclei with previous records of sarcoptic mange (affected nuclei, AN1-10), while the remaining three population nuclei were historically free of sarcoptic mange (HFN1-3) (Figure 1).

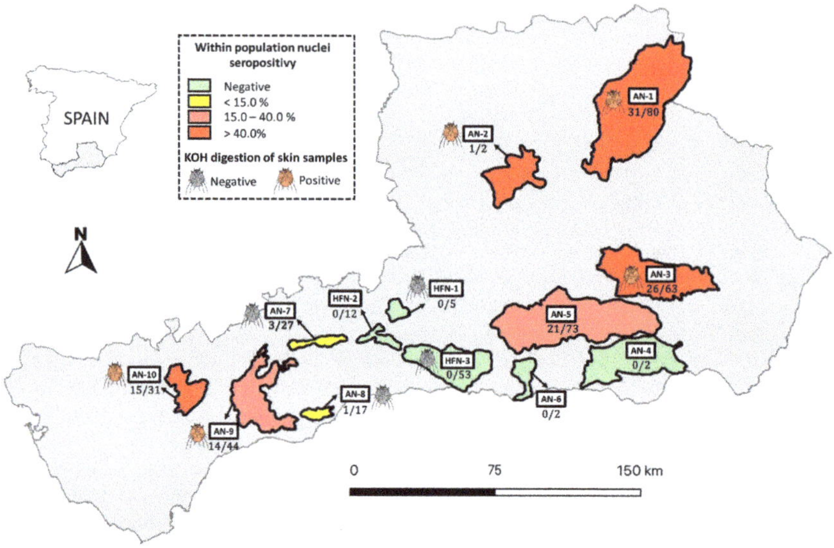

Figure 1. Spatial distribution of the Iberian ibex population nuclei sampled. AN: population nuclei with previous records of sarcoptic mange; HFN: population nuclei historically free of sarcoptic mange. Colour gradation shows within-population nuclei seropositivity. Fractions represent the number of ELISA-positive animals divided by the overall number of animals tested.

Sampled animals were classified according to the visual diagnosis of sarcoptic mange and included 157 individuals exhibiting external lesions compatible with sarcoptic mange (alopecia, scales, crusts, seborrhoea, hyperkeratosis, and/or skin lichenification, according to Pérez et al. [20]) and 254 clinically healthy Iberian ibexes (189 from AN and 65 from HFN) (Figure 2). Skin samples were collected from 88 of the 157 ibexes showing mange-compatible lesions. Animals with mange-compatible lesions were divided into four categories according to the percentage of skin surface area affected: grade I (≤25%); grade II, (25–50%); grade III (50–75%); and grade IV (≥75%) [20,21]. Data on location, age (juveniles: <2 years old; sub-adult: 2–6 years old; adult: >6 years old), sex, and date of sampling were recorded in all sampled individuals whenever possible (Table 1).

Figure 2. Adult male Iberian ibexes. (**A**) Individual with presence of external mange-compatible lesions. (**B**) Clinically healthy individual.

Table 1. Frequency of antibodies against *S. scabiei* in Iberian ibexes in Andalusia (southern Spain) and results of bivariate analysis.

Variable	Categories	Iberian Ibexes with Skin Lesions Compatible with Sarcoptic Mange			Clinically Healthy Iberian Ibexes		
		% ELISA Positive	Seropositives/ Overall [a]	*p*-Value	% ELISA Positive	Seropositives/ Overall [a]	*p*-Value
Location [b]	AN	68.4	104/152	0.004	4.2	8/189	0.09
	HFN	0.0	0/5		0.0	0/65	
Age	Juvenile	100.0	7/7	0.152	10.0	1/10	0.305
	Sub-adult	67.1	49/73		4.5	4/88	
	Adult	63.9	46/72		2.1	3/140	
Sex	Male	68.1	77/113	0.179	3.7	7/187	0.366
	Female	58.5	24/41		1.6	1/62	
Sampling year	2015	-	-	0.153	20.0	1/5	0.231
	2016	0.0	0/3		2.8	1/36	
	2017	60.6	20/33		1.7	1/58	
	2018	73.3	11/15		6.8	3/44	
	2019	66.1	39/59		1.5	1/66	
	2020	80.0	12/15		3.4	1/29	
	2021	68.8	22/32		0.0	0/16	
Percentage of skin surface area affected	Grade I (≤25%)	51.1	23/45	0.002	-	-	-
	Grade II (25–50%)	76.3	29/38				
	Grade III (50–75%)	83.3	10/12				
	Grade IV (≥75%)	88.2	30/34				

[a] Missing values omitted. [b] AN: population nuclei with previous records of sarcoptic mange; HFN: population nuclei historically free of sarcoptic mange.

In addition, cross-sectional sampling was carried out in domestic goats in the same period and study region. The sample size was calculated based on an estimated prevalence of 50% (which provides the highest sample size in studies with unknown prevalence) with a 95% confidence interval (CI95%) and a desired precision of ±5%, resulting in 385 specimens to be sampled. Goat flocks were randomly selected in the four provinces with the highest goat census [22] and included three of the five provinces where Iberian ibexes were sampled. Ultimately, 392 blood samples were randomly collected from 28 goat herds.

2.2. Serological Analyses

Blood samples from Iberian ibexes and domestic goats were collected through jugular vein puncture using sterile tubes without anticoagulants in live animals, or by puncture of the endocranial venous sinuses in dead Iberian ibexes, as previously described. After centrifugation, sera were collected and stored at −20 °C until analysis. The presence of antibodies against *S. scabiei* was determined using an in-house indirect ELISA, following the protocol described by Rambozzi et al. [23] but using commercial ELISA plates coated with *S. scabiei* var. *suis* antigen (Sarcoptes-Elisa 2001® PIG, AFOSA GmbH, Blankenfelde-Mahlow, Germany) [24]. The cut-off value used was 92.0% [24]. The estimated sensitivity (Se) and specificity (Sp) of this in-house ELISA were 93.0% and 93.5%, respectively [24].

2.3. *Potassium Hydroxide (KOH) Digestion Procedure*

Skin scrapings were obtained from lesions compatible with mange, encompassing both healthy and injured tissue, and were processed in a 10% KOH solution for 60 min at 37 °C. Subsequently, they were observed under the microscope (20× and 40×) for the detection of *S. scabiei*. Identification of mites was performed according to the keys and descriptions of Wall and Shearer [25].

2.4. *Statistical Analyses*

The frequency of seropositivity was estimated from the ratio of positive samples to the total number of samples analysed. The confidence intervals for seroprevalences were estimated by the standard error 95% confidence interval. Associations between the serological results and independent variables (location, age, sex, sampling year, and percentage of skin surface area affected) were analysed using Pearson's chi-squared test or Fisher's exact test, as appropriate. Pearson's chi-squared test evaluates the observed frequencies from a sample against the expected frequencies, assuming that there is no association between the variables and that the distribution is normal. For the test to be valid, it is essential that all expected frequencies are sufficiently large (>5). Therefore, when there were fewer than six observations per category, Fisher's exact test was used [26]. Differences were considered statistically significant when p-value < 0.05. Statistical analysis to determine significant differences between the serological results and independent variables was carried out using the SPSS statistical software package, version 25.0 (IBM Corporation, Somers, NY, USA).

3. Results

Antibodies against *S. scabiei* were detected in 112 (27.3 ± 4.3%) of the 411 Iberian ibexes analysed. Seropositivity was found in 104 of the 157 (66.2 ± 7.4%) animals with mange-compatible lesions, and in 8 of the 254 (3.1 ± 2.2%) clinically healthy ibexes. Out of 88 analysed skin scrapings, mites of *S. scabiei* were confirmed in 57 (64.8%) cases. Antibodies against *S. scabiei* were found in the sera of 49 (86.0 ± 9.0%) of these 57 animals.

At least one seropositive Iberian ibex was detected in 8 of the 13 (61.5%) population nuclei, with seropositive values ranging between 5.8% and 50%. Seropositive animals were detected in AN but not in HFN (Figure 1). Anti-*S. scabiei* antibodies were not detected in any of the 392 domestic goats. Sarcoptic mange-compatible lesions were not observed by farmers in the 28 goat flocks sampled.

Significant differences in seropositivity were observed among Iberian ibexes with mange-compatible lesions, depending on the affected skin surface. The frequency of antibodies significantly increased with the percentage of skin surface affected, as follows: 51.1 ± 14.6% (23/45) with grade I; 76.3 ± 13.5% (29/38) with grade II; 83.3 ± 21.1% (10/12) with grade III; and 88.2 ± 10.8% (30/34) with grade IV (p = 0.002) (Table 1). A temporal pattern was also observed, with a significantly higher seropositivity in spring (42.9 ± 8.2%; 60/140) compared to summer (23.7 ± 13.5%; 9/38; p = 0.023), autumn (8.8 ± 5.2%; 10/114; p < 0.001), and winter (27.7 ± 8.0%; 33/119; p = 0.008).

4. Discussion

Even though different direct and indirect methods have been proposed to investigate the circulation of *S. scabiei* in wild and domestic species—including clinical diagnosis, dermatoscopy, intradermal skin tests, infrared thermal imaging, antibody and antigen detection, PCR-based methods, or even the use of mange-detector dogs [5,27–29]—the diagnosis of sarcoptic mange remains a challenge today. Currently, there are few diagnostic methods showing satisfactory performance and cost–benefit balance [30]. Among them, indirect ELISA is deemed a valuable diagnostic tool for assessing *S. scabiei* exposure [2,24,30].

The overall individual prevalence of antibodies against *S. scabiei* detected in the Iberian ibex (27.3%) during the present study is evidence of a high ongoing circulation of the parasite in southern Spain. However, given that the analysed animals were opportunistically sampled in the framework of the epidemiological surveillance program, this result may eventually overestimate the actual prevalence of sarcoptic mange among them, thus, the outcome should be interpreted taking into account the different epidemiological contexts. Moreover, these results should be interpreted in the context of a broad time frame, taking into account that the risk factors could change over the time. Of the Iberian ibexes exhibiting mange-compatible lesions, 66.2% tested positive for anti-*S. scabiei* antibodies. This result is mostly explained by the lower Sp (60.7%) of the visual diagnosis of sarcoptic mange in Iberian ibexes [31] compared to the ELISA (93.5%) [24]. This interpretation is supported by the higher seroprevalence (86.0%) found in ibexes that exhibited mites of *S. scabiei* in their skin samples. However, eight cases of mange confirmed by KOH digestion were seronegative, confirming the estimated Se value of the ELISA test used. Complementarily, the onset of clinical dermatological signs following *S. scabiei* infestation may shortly anticipate the onset of a measurable humoral response, as shown by experimental trials in the Iberian ibex [24,32] and in other mammal species [33–36]. Consistently, within the group of animals showing mange-compatible lesions, we found a significant positive association between the presence of anti-*S. scabiei* antibodies and the affected skin surface—a reasonable proxy of the mite population size and the related antigenic stimulation of the host's immune system [20]. These findings suggest a cumulative persistence of antibodies throughout the course of the disease; moreover, they are further evidence of the poor defensive contribution of the circulating antibody response in the face of *Sarcoptes* invasion [2,18,37].

Interestingly, 3.1% of the clinically healthy Iberian ibexes showed antibodies against *S. scabiei*. This result could be associated to the Sp of the ELISA (93.5%) or may mirror alternative situations, such as (i) recovery after infection and the transitory development of mild lesions [15,38,39], or (ii) the presence of the disease at an early stage. Concerning the second hypothesis, although the Se of the visual inspection has been shown to be high (87.1%) [31], the presence of undetected skin lesions during the external inspection cannot be ruled out, particularly in recently infected animals. Additional studies are needed to evaluate these hypotheses.

A significantly higher seropositivity was observed in animals sampled in spring. This result is consistent with the higher number of cases of sarcoptic mange reported in Iberian ibexes during the winter season [12,27,31,40], as well as the subsequent seroconversion several weeks after infestation [32]. The cold period is particularly favourable for mite direct transmission because of the timing of the rutting season [11]. This temporal pattern can be also explained by the highest survival time of *S. scabiei* at <20 °C and relative humidity > 75% [2], which increases the risk of environmental (indirect) transmission during the winter period.

The detection of at least one seropositive animal in eight of the thirteen (61.5%) sampled population nuclei indicates that *S. scabiei* is widespread in this species in southern Spain. Accordingly, Fernández-Muñoz et al. [14] observed that sarcoptic mange is rapidly spreading among Iberian ibex populations across the Iberian Peninsula, particularly in the Mediterranean Basin. In our study, all seropositive individuals originated from AN, whereas seropositivity was not found in the Iberian ibexes from the three HFN analysed. Of note, the number of analysed sera from the remaining two nuclei was very limited (two samples each, Figure 1). From a conservation perspective, we highlight an interest in confirming, with a diagnostic method complementary to the traditional ones (e.g., observation from a distance), that no circulation of the parasite occurred in selected areas during the study period, despite proximity and likely connections with AN. The reasons why these nuclei have been mange-free for decades are unknown and undoubtedly worth investigating. Resilience in naïve nuclei of Iberian ibex and other caprine hosts to the severe demographic effects of sarcoptic mange has been already reported [8,15,41] but an authentic innate resistance would be unprecedented. If they exist, innately resistant nuclei should represent a favourite source of founder individuals for (i) planned reintroductions in areas connected with AN, and (ii) the restocking of those ibex nuclei which are still very sensitive to mange effects after decades of *Sarcoptes* circulation.

Since the first epizootic outbreak of sarcoptic mange was described in Iberian ibex populations in late 1987 [12], different studies have pointed out the relevance of multi-host systems in the transmission and maintenance of *S. scabiei* [4]. In this sense, transmission of *S. scabiei* at the wild Caprinae–domestic goat interface has been well documented [7,12,42,43]. In our study, none of the domestic goats analysed showed antibodies against this parasite, suggesting that sarcoptic mange is now maintained in an independent wild cycle in the study area. In line with this, Falconi et al. [44] observed that *S. scabiei* is self-maintained independently in Cantabrian chamois (*Rupicapra pyrenaica parva*) and livestock in northern Spain.

5. Conclusions

To the best of the authors' knowledge, this is the first seroepidemiological study that jointly assesses the exposure of *S. scabiei* at the Iberian ibex–domestic goat interface. Our results indicate a widespread distribution of this parasite among the Iberian Ibex populations of southern Spain and a null or limited circulation in domestic goat flocks from this region, suggesting that *S. scabiei* is maintained in exposed ibex nuclei, and in sympatric livestock if justified by the epidemiological context. Surveillance should inform the management of exposed ibex nuclei, including options such as selective culling of mangy individuals, depopulation, mass treatment, and *laissez-faire* [4,6]. Within this frame, the results of our survey suggest that sero-diagnosis by indirect ELISA represents a useful complementary tool to monitor the exposure of wild and domestic Caprinae to *S. scabiei* over time.

Author Contributions: Conceptualization, F.G.-G., A.A. and I.G.-B.; methodology, D.J.-M., D.D., D.C.-T. and B.M.; software, D.J.-M., D.C.-T. and I.G.-B.; validation, D.D., L.R., A.A. and I.G.-B.; formal analysis, D.J.-M., D.D., D.C.-T. and I.G.-B.; investigation, F.G.-G., D.J.-M., A.A., L.C.-S. and B.M.; resources, F.G.-G., L.R., C.V.C., L.C.-S. and I.G.-B.; data curation, D.J.-M. and D.C.-T.; writing—original draft preparation, F.G.-G. and D.J.-M.; writing—review and editing, F.G.-G., D.J.-M., A.A., L.R., C.V.C., L.C.-S., B.M., D.C.-T. and I.G.-B.; visualization, A.A. and I.G.-B.; supervision, L.R., D.C.-T. and I.G.-B.; project administration, I.G.-B.; funding acquisition, F.G.-G., L.R., C.V.C. and I.G.-B. All authors have read and agreed to the published version of the manuscript.

Funding: This research was partially funded by CIBER—Consorcio Centro de Investigación Biomédica en Red (CB 2021), Instituto de Salud Carlos III, Ministerio de Ciencia e Innovación, and Unión Europea—Next Generation EU. D. Jiménez-Martín was supported by an FPU grant (FPU22/03649) funded by the Spanish Ministry of Universities.

Institutional Review Board Statement: This study did not involve purposeful killing of animals. The collection of blood samples from Iberian ibex, captured alive, was part of the official Management Program of the Iberian Ibex in Andalusia, Spain. Samples from dead ibexes were collected from legally hunted animals, with the correct permits and licenses. Samples from domestic goats were collected during the official Animal Health Campaigns of the Regional Government of Andalusia, Spain. Therefore, no ethical approval was necessary.

Informed Consent Statement: Not applicable.

Data Availability Statement: The data presented in this study are available on request from the corresponding author.

Acknowledgments: We would like to thank everyone involved in the Epidemiological Surveillance Program in Wildlife and the Management Program of the Iberian Ibex in Andalusia for the collection of the samples. We are grateful to Manolo Sánchez for providing the photos used in Figure 2.

Conflicts of Interest: None of the authors of this study have a financial or personal relationship with other people or organizations that could inappropriately influence or bias the content of this paper.

References

1. Colebrook, E.; Wall, R. Ectoparasites of livestock in Europe and the Mediterranean region. *Vet. Parasitol.* **2004**, *120*, 251–274. [CrossRef] [PubMed]
2. Arlian, L.G.; Morgan, M.S. A review of *Sarcoptes scabiei*: Past, present and future. *Parasites Vectors* **2017**, *10*, 297. [CrossRef] [PubMed]
3. Escobar, L.E.; Carver, S.; Cross, P.C.; Rossi, L.; Almberg, E.S.; Yabsley, M.J.; Niedringhaus, K.D.; Van Wick, P.; Dominguez-Villegas, E.; Gakuya, F.; et al. Sarcoptic mange: An emerging panzootic in wildlife. *Transbound. Emerg. Dis.* **2022**, *69*, 927–942. [CrossRef] [PubMed]
4. Moroni, B.; Angelone, S.; Pérez, J.M.; Molinar Min, A.R.; Pasquetti, M.; Tizzani, P.; López-Olvera, J.R.; Valldeperes, M.; Granados, J.E.; Lavín, S.; et al. Sarcoptic mange in wild ruminants in Spain: Solving the epidemiological enigma using microsatellite markers. *Parasit. Vectors* **2021**, *14*, 171. [CrossRef]
5. Walton, S.F.; Currie, B.J. Problems in diagnosing scabies, a global disease in human and animal populations. *Clin. Microbiol. Rev.* **2007**, *20*, 268–279. [CrossRef] [PubMed]
6. Pérez, J.M.; Granados, J.E.; Espinosa, J.; Ráez-Bravo, A.; López-Olvera, J.R.; Rossi, L.; Meneguz, P.G.; Angelone, S.; Fandos, P.; Soriguer, R.C. Biology and management of sarcoptic mange in wild Caprinae populations. *Mamm. Rev.* **2021**, *51*, 82–94. [CrossRef]
7. Rossi, L.; Tizzani, P.; Rambozzi, L.; Moroni, B.; Meneguz, P.G. Sanitary emergencies at the wild/domestic Caprines interface in Europe. *Animals* **2019**, *9*, 922. [CrossRef] [PubMed]
8. Turchetto, S.; Obber, F.; Rossi, L.; D'Amelio, S.; Cavallero, S.; Poli, A.; Parisi, F.; Lanfranchi, P.; Ferrari, N.; Dellamaria, D.; et al. Sarcoptic Mange in Wild Caprinae of the Alps: Could Pathology Help in Filling the Gaps in Knowledge? *Front. Vet. Sci.* **2020**, *7*, 193. [CrossRef]
9. Cabrera, A. The Subspecies of the Spanish Ibex. *Proc. Zool. Soc. Lond.* **1911**, *81*, 963–977. [CrossRef]
10. Herrero, J.; Acevedo, P.; Arnal, M.C.; Fernández de Luco, D.; Fonseca, C.; García-González, R.; Pérez, J.M.; Sourp, E. *Capra pyrenaica*. The IUCN Red List of Threatened Species 2021: e.T3798A195855497. 2021. Available online: https://dx.doi.org/10.2305/IUCN.UK.2021-1.RLTS.T3798A195855497.en (accessed on 2 April 2024).
11. Fandos, P. *La cabra montés (Capra pyrenaica) en el Parque Natural de las Sierras de Cazorla, Segura y las Villas*; Serie Técnica ICONA: Madrid, Spain, 1991; ISBN 9788485496907.
12. León-Vizcaíno, L.; Ruíz de Ybáñez, M.R.; Cubero, M.J.; Ortíz, J.M.; Espinosa, J.; Pérez, L.; Simón, M.A.; Alonso, F. Sarcoptic mange in Spanish ibex from Spain. *J. Wildl. Dis.* **1999**, *35*, 647–659. [CrossRef]
13. Granados, J.E.; Ros-Candeira, A.; Pérez-Luque, A.J.; Moreno-Llorca, R.; Cano-Manuel, F.J.; Fandos, P.; Soriguer, R.C.; Cerrato, J.E.; Jiménez, J.M.P.; Ramos, B.; et al. Long-Term Monitoring of the Iberian Ibex Population in the Sierra Nevada of the Southeast Iberian Peninsula. *Sci. Data* **2020**, *7*, 203. [CrossRef] [PubMed]
14. Fernández-Muñoz, M.J.; Castillo-Contreras, R.; Pérez, J.M.; Granados, J.E.; Márquez, F.J.; López-Montoya, A.J. Co-infection patterns in the ectoparasitic community affecting the Iberian ibex *Capra pyrenaica*. *Parasit. Vectors* **2023**, *16*, 172. [CrossRef] [PubMed]
15. Valldeperes, M.; Yerro, P.P.; López-Olvera, J.R.; Fandos, P.; Lavín, S.; Escofet, R.C.S.; Mentaberre, G.; León, F.J.C.-M.; Espinosa, J.; Ráez-Bravo, A.; et al. Diseases of Iberian ibex (*Capra pyrenaica*). *Eur. J. Wildl. Res.* **2023**, *69*, 63. [CrossRef] [PubMed]
16. Bornstein, S.; Mörner, T.; Samuel, W.M. Sarcoptes scabiei and Sarcoptic Mange. In *Parasitic Diseases of Wild Mammals*; Iowa State University Press: Ames, IA, USA, 2001; pp. 107–119.
17. Pence, D.B.; Ueckermann, E. Sarcoptic manage in wildlife. *Rev. Sci. Tech. (Int. Off. Epizoot.)* **2002**, *21*, 385–398. [CrossRef]
18. Pérez, J.M.; Molina, L.; Ureña-Gutiérrez, B.; Espinosa, J.; López-Montoya, A.J.; Boos, M.; Granados, J.E.; Cano-Manuel, F.J.; Azorit, C. Individual stress responses to *Sarcoptes scabiei* infestation in Iberian ibex, *Capra pyrenaica*. *Gen. Comp. Endocrinol.* **2019**, *281*, 1–6. [CrossRef] [PubMed]
19. CSMAEA. Consejería de Sostenibilidad, Medio Ambiente y Economía Azul. Junta de Andalucía. Available online: https://www.juntadeandalucia.es/medioambiente/portal/documents/20151/1487460/pacam_web_2013.pdf/131a3a48-0885-b3a2-ff5e-a58d0a1d2c19 (accessed on 2 February 2024).
20. Pérez, J.M.; Granados, J.E.; Sarasa, M.; Serrano, E. Usefulness of estimated surface area of damaged skin as a proxy of mite load in the monitoring of sarcoptic mange in free-ranging populations of Iberian wild goat, *Capra pyrenaica*. *Vet. Parasitol.* **2011**, *176*, 258–264. [CrossRef] [PubMed]
21. Ráez-Bravo, A.; Granados, J.E.; Cerón, J.J.; Cano-Manuel, F.J.; Fandos, P.; Pérez, J.M.; Espinosa, J.; Soriguer, R.C.; López-Olvera, J.R. Acute phase proteins increase with sarcoptic mange status and severity in Iberian ibex (*Capra pyrenaica*, Schinz 1838). *Parasitol. Res.* **2015**, *114*, 4005–4010. [CrossRef]
22. MAPA. Ministerio de Agricultura Pesca y Alimentación. Available online: https://www.mapa.gob.es/es/estadistica/temas/estadisticas-agrarias/resultados_definitivos_nov17_ovino-caprino_webmapama_tcm30-450864.pdf (accessed on 6 March 2024).
23. Rambozzi, L.; Menzano, A.; Lavin, S.; Rossi, L. Biotin-avidin amplified ELISA for detection of antibodies to Sarcoptes scabiei in chamois (*Rupicapra* spp.). *Vet. Res.* **2004**, *35*, 701–708. [CrossRef] [PubMed]

24. Ráez-Bravo, A.; Granados, J.E.; Serrano, E.; Dellamaria, D.; Casais, R.; Rossi, L.; Puigdemont, A.; Cano-Manuel, F.J.; Fandos, P.; Pérez, J.M.; et al. Evaluation of three enzyme-linked immunosorbent assays for sarcoptic mange diagnosis and assessment in the Iberian ibex, *Capra pyrenaica*. *Parasit. Vectors* **2016**, *9*, 558. [CrossRef]
25. Wall, R.; Shearer, D. *Veterinary Entomology: Arthropod Ectoparasites of Veterinary Importance*; Springer Science & Business Media: Dordrecht, The Netherland, 1997; ISBN 9780412615108.
26. Thrusfield, M.; Christley, R. *Veterinary Epidemiology*; Wiley: Hoboken, NJ, USA, 2018; ISBN 9781118280287.
27. Arenas, A.J.; Gómez, F.; Salas, R.; Carrasco, P.; Borge, C.; Maldonado, A.; O'Brien, D.J.; Martínez Moreno, F.J. An evaluation of the application of infrared thermal imaging to the tele-diagnosis of sarcoptic mange in the Spanish ibex (*Capra pyrenaica*). *Vet. Parasitol.* **2002**, *109*, 111–117. [CrossRef]
28. Alasaad, S.; Permunian, R.; Gakuya, F.; Mutinda, M.; Soriguer, R.C.; Rossi, L. Sarcoptic-mange detector dogs used to identify infected animals during outbreaks in wildlife. *BMC Vet. Res.* **2012**, *8*, 110. [CrossRef]
29. Angelone-Alasaad, S.; Molinar Min, A.; Pasquetti, M.; Alagaili, A.N.; D'Amelio, S.; Berrilli, F.; Obanda, V.; Gebely, M.A.; Soriguer, R.C.; Rossi, L. Universal conventional and real-time PCR diagnosis tools for *Sarcoptes scabiei*. *Parasit. Vectors* **2015**, *8*, 587. [CrossRef] [PubMed]
30. Rampton, M.; Walton, S.F.; Holt, D.C.; Pasay, C.; Kelly, A.; Currie, B.J.; McCarthy, J.S.; Mounsey, K.E. Antibody responses to *Sarcoptes scabiei* apolipoprotein in a porcine model: Relevance to immunodiagnosis of recent infection. *PLoS ONE* **2013**, *8*, e65354. [CrossRef] [PubMed]
31. Valldeperes, M.; Granados, J.E.; Pérez, J.M.; Castro, I.; Ráez-Bravo, A.; Fandos, P.; López-Olvera, J.R.; Serrano, E.; Mentaberre, G. How sensitive and specific is the visual diagnosis of sarcoptic mange in free-ranging Iberian ibexes? *Parasit. Vectors* **2019**, *12*, 405. [CrossRef] [PubMed]
32. Sarasa, M.; Rambozzi, L.; Rossi, L.; Meneguz, P.G.; Serrano, E.; Granados, J.-E.; González, F.J.; Fandos, P.; Soriguer, R.C.; Gonzalez, G.; et al. *Sarcoptes scabiei*: Specific immune response to sarcoptic mange in the Iberian ibex *Capra pyrenaica* depends on previous exposure and sex. *Exp. Parasitol.* **2010**, *124*, 265–271. [CrossRef] [PubMed]
33. van der Heijden, H.M.J.F.; Rambags, P.G.M.; Elbers, A.R.W.; van Maanen, C.; Hunneman, W.A. Validation of ELISAs for the detection of antibodies to *Sarcoptes scabiei* in pig. *Vet. Parasitol.* **2000**, *89*, 95–107. [CrossRef] [PubMed]
34. Bornstein, S.; Zakrisson, G.; Thebo, P. Klinisk bild och antikroppssvar vid experimentell infektion av Sarcoptes scabiei var, vulpes hos rödräv (*Vulpes vulpes*). *Acta Vet. Scand.* **1995**, *36*, 509–519. [CrossRef] [PubMed]
35. Tarigan, S. Antibody response in naïve and sensitised goats infested by *Sarcoptes Scabiei*. *J. Ilmu Ternak Dan Vet.* **2014**, *9*, 258–265.
36. Casais, R.; Dalton, K.P.; Millán, J.; Balseiro, A.; Oleaga, Á.; Solano, P.; Goyache, F.; Prieto, J.M.; Parra, F. Primary and secondary experimental infestation of rabbits (*Oryctolagus cuniculus*) with *Sarcoptes scabiei* from a wild rabbit: Factors determining resistance to reinfestation. *Vet. Parasitol.* **2014**, *203*, 173–183. [CrossRef]
37. Castro, I.; de la Fuente, A.; Fandos, P.; Cano-Manuel, F.; Granados, J.-E.; Soriguer, R.C.; Alasaad, S.; Pérez, J.M. On the population biology of *Sarcoptes scabiei* infesting Iberian ibex (*Capra pyrenaica*). *Int. J. Acarol.* **2016**, *42*, 7–11. [CrossRef]
38. Niedringhaus, K.D.; Brown, J.D.; Ternent, M.A.; Peltier, S.K.; Van Wick, P.; Yabsley, M. Serology as a tool to investigate sarcoptic mange in American black bears (*Ursus americanus*). *J. Wildl. Dis.* **2019**, *56*, 350–358. [CrossRef] [PubMed]
39. Valldeperes, M.; Granados, J.E.; Pérez, V.; López-Olvera, J.R.; Ráez-Bravo, A.; Fandos, P.; Pérez, J.M.; Mentaberre, G.; Tampach, S.; Soriguer, R.C.; et al. The local skin cellular immune response determines the clinical outcome of sarcoptic mange in Iberian ibex (*Capra pyrenaica*). *Front. Vet. Sci.* **2023**, *10*, 1183304. [CrossRef] [PubMed]
40. Pérez, J.M.; Ruiz-Martínez, I.; Granados, J.E.; Soriguer, R.C.; Fandos, P. The dynamics of sarcoptic mange in the ibex population of Sierra Nevada in Spain—Influence of climatic factors. *J.Wild. Res.* **1997**, *2*, 86–89.
41. Rossi, L.; Fraquelli, C.; Vesco, U.; Permunian, R.; Sommavilla, G.M.; Carmignola, G.; Da Pozzo, R.; Meneguz, P.G. Descriptive epidemiology of a scabies epidemic in chamois in the Dolomite Alps, Italy. *Eur. J. Wildl. Res.* **2007**, *53*, 131–141. [CrossRef]
42. Fernández-Morán, J.; Gómez, S.; Ballesteros, F.; Quirós, P.; Benito, J.; Feliu, C.; Nieto, J. Epizootiology of sarcoptic mange in a population of cantabrian chamois (*Rupicapra pyrenaica parava*) in Northwestern Spain. *Vet. Parasitol.* **1997**, *73*, 163–171. [CrossRef] [PubMed]
43. Menzano, A.; Rambozzi, L.; Rossi, L. A severe episode of wildlife-derived scabies in domestic goats in Italy. *Small Rumin. Res.* **2007**, *70*, 154–158. [CrossRef]
44. Falconi, C.; Oleaga, Á.; López-Olvera, J.R.; Casais, R.; Prieto, M.; Gortázar, C. Prevalence of antibodies against selected agents shared between Cantabrian chamois (*Rupicapra pyrenaica parva*) and domestic goats. *Eur. J. Wildl. Res.* **2010**, *56*, 319–325. [CrossRef]

animals

MDPI

Article

Transmission Dynamics of Imported Vaccine-Origin PRRSV-2 within and between Commercial Swine Integrations in Hungary

Szilvia Jakab [1,2,†], Krisztián Bányai [1,2,3,*,†], Krisztina Bali [1,2], Imre Nemes [4], Ádám Bálint [5] and István Szabó [4,*]

[1] HUN-REN Veterinary Medical Research Institute, Hungária krt. 21., H-1143 Budapest, Hungary; jakab.szilvia@vmri.hu (S.J.); bali.krisztina@vmri.hu (K.B.)
[2] National Laboratory for Infectious Animal Diseases, Antimicrobial Resistance, Veterinary Public Health and Food Chain Safety, Hungária krt. 21., H-1143 Budapest, Hungary
[3] Department of Pharmacology and Toxicology, University of Veterinary Medicine, István utca. 2., H-1078 Budapest, Hungary
[4] National PRRS Eradication Committee, Keleti Károly utca. 24., H-1024 Budapest, Hungary; nemesi@nebih.gov.hu
[5] Veterinary Diagnostic Directorate, National Food Chain Safety Office, H-1143 Budapest, Hungary; balintad@nebih.gov.hu
* Correspondence: banyai.krisztian@vmri.hu (K.B.); iszabodr@t-online.hu (I.S.)
† These authors contributed equally to this work.

Simple Summary: Two recent transmission chains of PRRSV-2-associated cases were documented in this study using field epidemiological and molecular genetic tools. The investigation highlighted the risks associated with the free movement of livestock in the European Union. To minimize this risk of re-infection of PRRS-free herds with PRRSV through animal imports, it is recommended that pigs are transported directly from the exporting holdings without the involvement of transit stations. Alternatively, the transit stations could be converted so that pigs in transit avoid contact with each other, thus preventing exposure to PRRSV infection.

Abstract: This study reports on the molecular epidemiology of Ingelvac-PRRS-MLV-associated cases in Hungary for the period 2020–2021. Field epidemiology investigations led the experts to conclude that imported pigs, which were shipped through transit stations in Denmark, introduced the vaccine virus. The movement of fatteners and the neglect of disease control measures contributed to the spread of the virus to PRRS-free pig holdings in the vicinity. Deep sequencing was performed to genetically characterize the genes coding for the virion antigens (i.e., ORF2 through ORF7). The study isolates exhibited a range of 0.1 to 1.8% nucleotide sequence divergence from the Ingelvac PRRS MLV and identified numerous polymorphic sites (up to 57 sites) along the amplified 3.2 kilo base pair genomic region. Our findings confirm that some PRRSV-2 vaccine strains can accumulate very high number of point mutations within a short period in immunologically naive pig herds.

Keywords: NA-type PRRSV; molecular epidemiology; virus transmission; next-generation sequencing; single nucleotide variation; Ingelvac PRRS MLV; Hungary

1. Introduction

Porcine reproductive and respiratory syndrome (PRRS) is amongst the most significant infectious diseases constantly inducing great economic losses for the global swine production. The typical manifestations of the syndrome are reproductive disorder in sows and respiratory disease in piglets [1]. The increased mortality of prenatal, pre-weaned and young pigs as well as reduced growth rate have the most relevant economic impact [2,3].

The causative agent of PRRS is an enveloped, single-stranded RNA virus (PRRSV) belonging to the *Arteriviridae* family [4]. Two, distantly related PRRSV species are distinguished, such as *Betaarterivirus suid 1* and *Betaarterivirus suid 2* (former PRRSV-1 or EU-type and PRRSV-2 or NA-type, respectively) [5].

Recognized by the competent committee of the European Union, Hungary launched an eradication program (2014–) in order to reduce the initially high prevalence of PRRSV and to alleviate the associated economic burden [6]. By 2022, all pig herds became free of wild-type PRRSVs, and the legislation continues to support the maintenance of an absolute free status for pig holdings (including disease-free status from vaccine-origin PRRSVs) [7]. The extensive monitoring conducted throughout the years of disease elimination revealed that the circulating EU-type Hungarian PRRSV strains were genetically diverse, whereas NA-type PRRSVs were scarcely identified [8,9]. In fact, all NA-type viruses detected after 2014 were classified as derivatives of a live, attenuated vaccine virus [9]. The vaccine virus was developed from a wild-type virulent strain, VR2332 (https://bi-animalhealth.com/swine/products/flex/ingelvac-prrs-mlv (accessed on 29 September 2023.)). The vaccine (Ingelvac PRRS MLV or RespPRRS MLV) is registered in nine countries within the EU, such as Denmark, Belgium, Germany, Lithuania, Luxembourg, Netherlands, Poland, Portugal and Spain (https://www.ema.europa.eu/en/documents/referral/modified-live-porcine-respiratory-reproducti-article-35-referral-annex-i-ii-iii_en.pdf (accessed on 29 September 2023)). In addition, Slovakia permits the use of this vaccine for immunization on a case-by-case basis. Identification of vaccine-origin PRRSV-2 strains has been published from the EU [10–14], and the policies concerning the registration of the PRRSV-2 vaccines are likely to reflect the actual epidemiological situation in the respective countries; yet, updated surveillance data are not regularly published. Although the Ingelvac PRRS MLV is not authorized in Hungary, the introduction of the vaccine virus is plausible, for example, through illegal import of vaccines (e.g., as happened in 2008, I. Szabó, unpublished data) or with imported, freshly vaccinated piglets.

Denmark, where the Ingelvac PRRS MLV is regularly used, is a major source of pre-fatteners for Hungary, therefore it is essential that the Hungarian authorities are aware of the export procedures. Under Danish legislation, Hungary is a red zone country, therefore exported pigs must be loaded onto Hungarian trucks at EU-certified collecting stations, not directly at the sellers' farm. Given that the transit stations place pigs from various farms in shared air space, even a PRRS-free shipment could be infected with various PRRSVs before entering Hungary [15].

In this study, we present the molecular epidemiological aspects of two PRRS outbreaks associated with the Ingelvac PRRS MLV. We wish to draw attention to the mode of infection, which is related to the commercial logistics of immunized fattening pigs in a foreign country.

2. Materials and Methods

2.1. Epidemiological Investigation

Our epidemiological investigations in relation to the accumulation of NA-type PRRSV infections involved seven large-scale fattening and breeding farms (Farm A to Farm G) during 2020–2021 (Table 1). The inspection of official departments included the checking of import licenses, the documents accompanying the shipments (TRACES certificate), and the results of diagnostics. At the same time, information was collected on the movement of transport vehicles (live animals and carcasses, forages, breeding material, etc.) passing through other establishments or slaughterhouses linked to the farms concerned. Additionally, the personnel entering the examined pig farms were recorded.

Table 1. Farms involved in the transmission chain of Ingelvac PRRS MLV.

Farms	Farm Size and Type	Origin of Pigs	Other Information
Farm A	4000-seat large-scale fattening farm, continuously supplied	Prefatteners are purchased from Denmark	Situated in the same county
Farm B	600-seat large-scale breeding farm (Hungarian Large White x Hungarian Landrace local hybrid), farrow-to-finish type establishment		
Farm C	large-scale breeding farm (1820 sow, Danbred hybrid), farrow-to-finish type establishment	Parts of prefatteners are transported to a separate farm or other integrators	
Farm D	2700-seat large-scale fattening farm, continuously supplied	Prefatteners are from Farm C	
Farm E	9000-seat large-scale fattening farm, continuously supplied	Prefatteners are from Farm C	-
Farm F	2000-seat large-scale fattening farm continuously supplied	Prefatteners are purchased from Denmark	-
Farm G	2600-seat large-scale fattening farm continuously supplied	Prefatteners are purchased from Denmark	-

2.2. PRRSV Diagnostics

Serum or buccal swab samples were collected from imported prefattener stocks, from pigs showing clinical symptoms and from pigs involved in regulatory PRRSV monitoring (pregnant gilts, sows and fatteners). In general, the number of samples taken for diagnostics is calculated based on 95% confidence at 10% prevalence values [16]. The viral RNA was isolated with the QIAmp Viral RNA Mini Kit (Qiagen, Hilden, Germany) and tested for PRRSV with RT-qPCR using the Virotype PRRSV RT-PCR Kit (Qiagen, Hilden, Germany) according to the manufacturer's instructions. Sequencing of ORF5 or ORF7 was performed on randomly selected samples exhibiting low Ct values obtained in the diagnostic RT-qPCR assay (for details see [17]). Sequencing of these amplified regions using the Sanger method was carried out on an ABI PRISM 3100 automatic sequencer. Antibody detection was performed using the PRRS Universal ELISA Kit (Ingenasa, Madrid, Spain) according to the manufacturer's instructions.

Concerning the imported fattening stocks, the measures initiated by the eradication program in accordance with the law are as follows. The entering prefatteners are quarantined for 60 days, and diagnostic tests are conducted (PRRSV RNA and antibody detection) 48 h within their arrival. At the end of the quarantine, the whole workflow is repeated. In addition to the monitoring of imported fattening stocks, regular and mandatory surveillance is performed every six months as required by the law [18].

2.3. Analysis of the ORF2-7 Region

The standard diagnostic procedures were supplemented with amplification and next-generation sequencing (NGS) of the structural ORFs, ORF2 to ORF7. In brief, the viral RNA was freshly re-isolated from the remaining specimens with the QIAmp Viral RNA Mini Kit (Qiagen, Hilden, Germany). SuperScript™ III Reverse Transcriptase (Invitrogen, Waltham, MA, USA) and an anchored poly(dT) primer (5′-TTTTTTTTTTTTTTTTTTAATTWCG-3′) were used for cDNA synthesis. Amplification of the ORF2-7 region was performed with the Phusion DNA polymerase (Thermo Scientific, Waltham, MA, USA). The final volume of PCR mixture was 20 µL, in 1× HF Buffer with 3% DMSO, 0.5 µL of each primer (10 µM), and 1 µL of cDNA template. The degenerate primer pair was as follows: 5′-CGKGCGCGCCAGRAAGGGAAAATTTA-3′ and 5′-GCACARTRTCAATCAGTGCCATTCAC-3′. The parameters of the PCR program were as follows: initial denaturation at 98 °C for 30 s, then 35 cycles of denaturation at 98 °C for 10 s, annealing at 66 °C for 30 s and extension at 72 °C for 2 min; the final elongation at

72 °C lasted for 10 min. After analysing the PCR product on 1% agarose gel, the bands were excised and purified with Gel/PCR DNA Fragments Extraction Kit (Geneaid, New Taipei City, Taiwan). For library preparation, we used the Nextera XT DNA Library Preparation Kit and the Nextera XT Index Kit v2 (Illumina, San Diego, CA, USA). Sequencing was performed on an Illumina® NextSeq 500 sequencer, as described previously [19]. GenBank accession numbers of the study strains are as follows: OR143097-OR143102.

2.4. Sequence Analysis

The whole ORF2-7 region was assembled from the NGS short reads via reference mapping to the VR2332 prototype genome. Variant calling (minimum variant frequency of 10%) was performed in Geneious 9.1.8 software (Biomatters, Inc., Auckland, New Zealand). The final consensus sequences were aligned with selected lineage-specific sequences using the MAFT algorithm as implemented in Geneious 9.1.8. Pairwise sequence identities were generated in the MEGA software (version 10.1.8) [20]. Phylogenetic reconstruction based on maximum likelihood was performed with MEGA software (version 10.1.8) (model HKY + G, 1000 bootstrap).

3. Results

3.1. Transmission Chains

Preceding the incidents, all affected pig herds in this study possessed a PRRS-free status registered by the Hungarian Veterinary Authorities. However, from autumn 2020 to summer 2021, two distinct transmission chains of the Ingelvac PRRS MLV vaccine virus were identified. In this section, we describe the timeline of sampling and diagnostic test results followed by the traditional and molecular epidemiologic findings.

3.1.1. Transmission Chain I

All identified cases in the transmission chain I were connected to the collecting station "P" in Denmark.

Farm A: Several shipments arrived from a foreign, certified PRRS-free farm to Farm A through the same collecting station in Denmark ("P") on each week in October and twice in the first two weeks of November 2020. In mid-October, 30 samples were subjected to laboratory tests from the first two stocks; however, that was several days after the mandatory 48 h. All samples had a positive PCR result, 20 samples also gave positive ELISA and a single PRRS EU strain (Porcilis PRRS vaccine strain) was detected via ORF5 sequencing. Neither relevant blood test nor ELISA were performed for the subsequent stocks.

Farm B: Routine laboratory tests of serum samples (n = 33) from Farm B were positive for PRRS (39% PCR positive and 100% seropositive) on 18 November 2020. The ORF7 sequence determined for a single sample (sample ID.: 64196) showed the closest genetic similarity to Ingelvac PRRS MLV.

Farm C: On 15 and 20 April 2021, sera (n = 11) and buccal swabs (n = 8) were collected from sows, respectively, which manifested mild clinical symptoms such as loss of appetite, fever and snorting. PRRS infection was confirmed with positive ELISA (100%) and PCR tests (100%). The ORF7 and ORF5 sequences (sample ID.: 18601, sow, buccal swab) identified the Ingelvac PRRS MLV as the causative PRRSV strain. Subsequently, the authorities ordered herd closure and monitoring of all age groups, and serum samples (n = 190) were gathered on 21 April. The obtained seropositivity and positive PCR rate (found in the cohorts of 28- and 60-day old pigs) for the corresponding herd were 74% and 7.9%, respectively. The ORF7 and ORF5 sequences (samples ID.: 19001, pig, serum) were classified as Ingelvac PRRS MLV.

Farm D: Fattening stocks were continuously placed on Farm D from Farm C. On 21 April 2021, routine serological tests showed 100% seropositivity and 16.7% of serum samples (n = 30) were PCR-positive in case of the fatteners (90–180 days old) settled in January and February of 2021. Sequencing of the ORF7 (samples ID.: 19702) resulted in the detection of Ingelvac PRRS MLV. Furthermore, seropositive animals were found among pigs settled in April 2021.

Farm E: Fattening stocks were continuously placed on Farm E from Farm C. On 23 April 2021, routine serological tests from serum samples (n = 160) showed almost 100% seropositivity and PCR positivity (1.9%) of the examined herd of Farm E, and Ingelvac PRRS MLV was identified after sequencing the ORF5 gene (sample ID.: 19601).

3.1.2. Transmission Chain II

All identified cases in the transmission chain II were connected to the transit station "T" in Denmark.

Farm F: Prefatteners purchased from a foreign-certified PRRS-free farm were imported through the transit station "T" in Denmark to Farm F on 26 May 2021. The mandatory laboratory tests from serum samples (n = 29) performed immediately after unloading showed high rates of positivity with PCR (100%) and ELISA (13.8%). The following diagnostic tests (7, 14 and 27 days after the settlement) confirmed the infection with PRRSV, and the Ingelvac PRRS MLV was identified via sequencing of the ORF7 (sample ID.: 27392).

Farm G: Prefatteners were ordered from a foreign-certified PRRS-free farm and transported through the collecting station "T" in Denmark to Farm G on 4 May 2021. Serology testing (n = 29) was conducted within 48 h after the arrival, which gave negative test results. At the end of the quarantine (7 July 2021); however, the herd displayed positivity with PCR (100%) and ELISA (100%) tests, and the pathogen was classified as the Ingelvac PRRS MLV based on ORF7 sequencing.

The events of pig import and the results of the PRRSV PCR tests conducted for this study are summarized on a timeline figure (Figure 1).

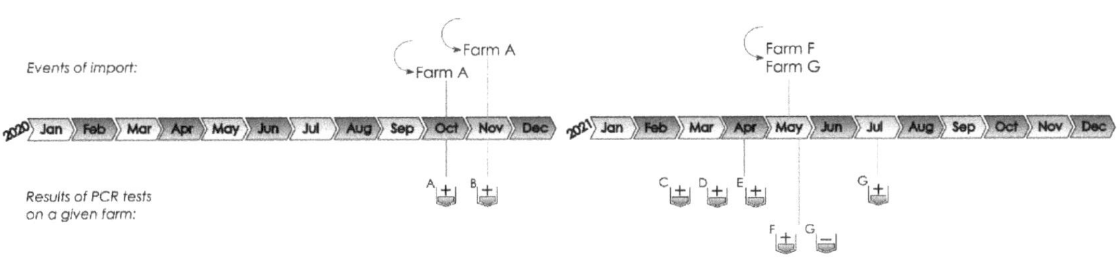

Figure 1. Schematic presentation of pig import events and the results of PRRSV PCR tests.

3.2. Epidemiological Investigation

Epidemiological investigation was performed to explore the most likely transmission route of the Ingelvac PRRS MLV vaccine strain among affected herds (Figure 2).

In the case series linked to collecting station "P" in Denmark, five large-scale farming units were involved, and Farm A was identified as the initial piece of the transmission chain. Pigs imported from Denmark through collecting station "P" to Farm A (October and November, 2020) could be infected recently with the Ingelvac PRRS MLV vaccine virus. From some stocks of Farm A, we identified the Porcilis PRRS vaccine strain, an EU-type vaccine that is regularly used for immunization in Denmark. The PRRSV-2 vaccine virus itself could not be detected at Farm A, in part, as a result of omitting routine virological testing of some imported stocks. Concerning the next steps in virus transmission, field investigation suggested that a manure suction vehicle, which regularly operated between Farm A and Farm B, was responsible for the introduction of PRRSV from Farm A to Farm B. Importantly, Farm B had no direct transport or any other commercial relationships with collecting station "P" or Farm A that could have explained the introduction of vaccine virus to this farm. Next, the authorities recognized that the route of infection to other farms could be epidemiologically linked to the management of fattener transport to the slaughterhouses or the pig movement between consecutive phases of production. An apparent issue was that weighing equipment was not available at Farm D; therefore, the transport vehicle owned by a local slaughterhouse was used to transport the pigs at to Farm C, where measuring equipment was available. Considering this practice, infection of Farms C and D with the vaccine strain may have occurred when the vehicle of the slaughterhouse collected infected pigs from Farm B on 17 February 2021, then the animals were weighed at Farm C. Next, the vehicle picked up fatteners at Farm D and returned to Farm C to weigh the pigs of Farm D. At last, the vehicle completed its journey at the slaughterhouse. Throughout the process of loading, the staff members had multiple contact events with infected animals, including the transport vehicle and many objects located at the farm that could have readily spread the vaccine virus. Another plausible explanation is that only Farm C was infected this way, and later, fatteners from Farm C were placed in Farm D, which caused the emergence of vaccine-associated PRRS on Farm D. In case of Farm E, the transportation of already infected prefatteners from Farm C to Farm E introduced the vaccine-origin PRRSV.

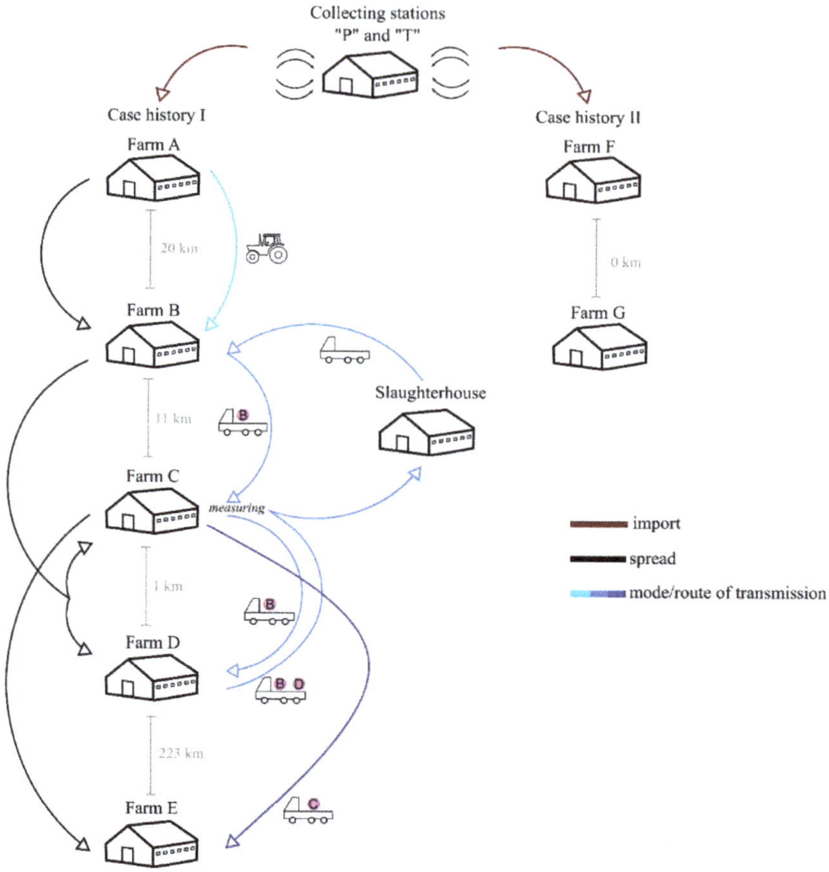

Figure 2. The most likely transmission routes of the Ingelvac PRRS MLV vaccine strain determined during field epidemiology investigation. The arrows indicate the movement of pigs and the possible routes of vaccine strain transmission.

In the case series linked to collecting station "T" in Denmark, two large-scale fattening holdings were exposed to the vaccine virus. A shipment to Farm F from Denmark very likely carried the PRRS vaccine strain, as diagnostic tests upon arrival identified severe PRRS infection in imported piglets. Farms F and G are two neighbouring establishments, with a fence separating them on the side boundary. The field investigation established that the fence and the negligible distance between the pens of the two farms could unlikely prevent the spread of vaccine virus to the PRRS-free herd of Farm G.

3.3. Molecular Investigations

Considering only the ORF5 and ORF7 regions, which were sequenced during the diagnostic workflow, the nucleotide (nt) and amino acid (aa) sequence identity among samples were 96.4% to 100% (nt) and 97.5% to 100% (as) for the ORF5, and 97.6% to 100% (nt) and 99.2% to 100% (as) for the ORF7. Both the ORF5 and ORF7 sequences shared high similarities to the Ingelvac PRRS MLV vaccine strain (98–100% and 98.1–100%), respectively. Thus, the sequence data obtained for ORF5/ORF7 uncovered the vaccine origin of strains in disease etiology; however, additional data were needed to support the field epidemiology observations. The genomic region encoding the structural proteins was amplified and sequenced from samples collected at Farms B, C, D, E and F. Unfortunately, some imported stocks having arrived at Farm A were not sampled, and amplification of the ORF2-7 region from samples collected at Farm G failed.

The ORF2-7 region was uniformly 3188 base pair long, neither indel mutations nor recombination events were detected. Pairwise nt identities of the obtained ORF2-7 consensus sequences fell between 97.5% and 100%, and, when compared to the Ingelvac PRRS MLV vaccine strain, the identities ranged between 98.2% and 99.9% (Table 2). By analysing the NGS runs, we found a total of 72 SNV sites in four samples (19001, 19702, 19601 and 27392) under the criterion of a minimum variant frequency of 10%. The other two samples (64196 and 18601) showed no sequence variation in the amplified genomic region. The frequency of SNVs, analysed at an average sequencing depth of 8976–22052X, ranged from 10.2% to 50.8% (Figure 3). The position of SNVs varied considerably among strains. Between the Ingelvac PRRS MLV and the VR2332 strain, the following nt positions were identified as being different along the ORF2-7 region: 610, 722, 813, 1967 and 3161. Only, two and one SNV variants identical with the VR2332 were detected in samples 19702 and 27392, respectively, at positions 610, 772 and 1967.

Table 2. Percent pairwise nucleotide identities among study strains and selected reference strains.

Strains	VR2332	Ingelvac PRRS MLV	64196 Farm_B	18601 Farm_C	19001 Farm_C	19702 Farm_D	19601 Farm_E	27392 Farm_F
VR2332								
Ingelvac PRRS MLV	99.6							
64196_Farm_B	98.8	99						
18601_Farm_C	98.8	99	100					
19001_Farm_C	98.1	98.3	97.6	97.6				
19702_Farm_D	99.6	99.9	99	99	98.3			
19601_Farm_E	98.1	98.2	97.5	97.5	99.7	98.2		
27392_Farm_F	99.3	99.4	98.8	98.8	98.3	99.5	98.2	

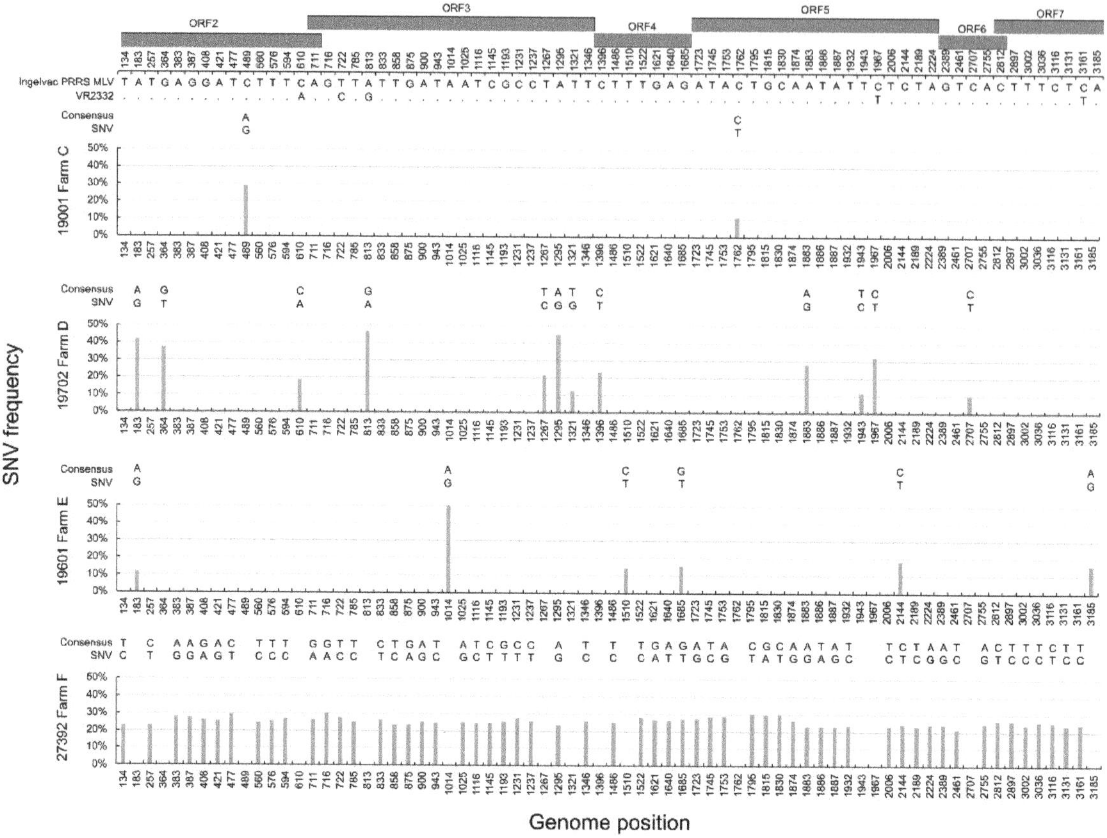

Figure 3. Distribution of SNV sites in the ORF2-7 region of study strains and their comparison to the consensus sequences of Ingelvac PRRS MLV and VR2332.

When analysing the aa identity values in the amplified structural protein region, the study strains were found to share 95.3–100% identities with the Ingelvac PRRS MLV vaccine strain, with GP3 being the most different region. In particular, the number of amino acid positions showing substitutions within GP2, GP3, GP4, GP5, M and N of the Hungarian isolates compared to the vaccine strain were the following: 11/256 (4.3%), 13/254 (5.1%), 3/178 (1.7%), 7/200 (3.5%), 3/174 (1.8%) and 1/123 (0.8%), respectively. As shown in Figure 4, the Hungarian study strains shared some unique aa residues (such as two ($F^{10} \rightarrow L^{10}$, $I^{237} \rightarrow M^{237}$), one ($T^{30} \rightarrow A^{30}$) and one ($D^{34} \rightarrow H/A/G^{34}$) aa residues in the GP2, GP3 and GP5, respectively) that were shared by the majority of study strains (with the exception of sample originating from Farm D) but differed from both the vaccine strain and its wild-type parental strain. We observed several aa residues that were identical to the Ingelvac PRRS MLV within GP2, GP3, GP5 and N, at three (H^{204}, A^{223} and V^{241}), one (I^{15}), two (C^{11}, A^{29}) and one (K^{10}) positions, respectively (Figure 5). On the contrary, only two aa residues situated in the GP3 ($I^{94} \rightarrow V^{94}$) and M ($E^{16} \rightarrow Q^{16}$), except 19702 from Farm D, were equal to the VR2332, respectively (Figure 5).

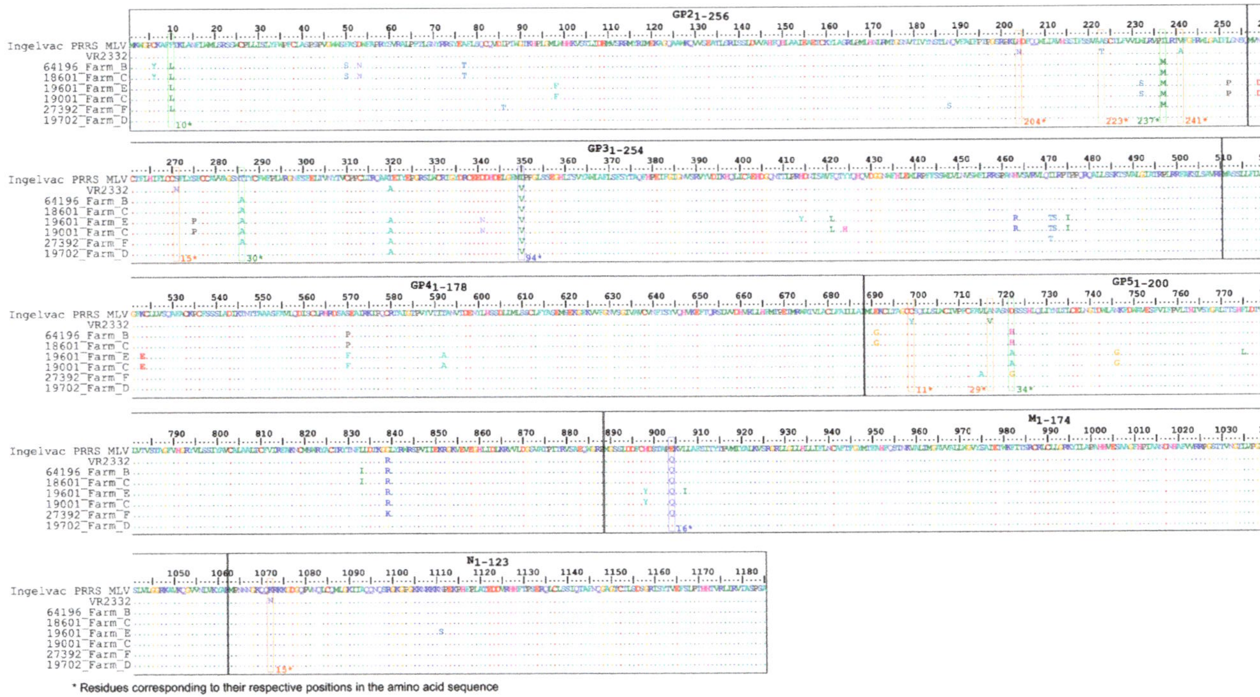

Figure 4. Concatenated amino acid alignment of the reference and Hungarian PRRSV-2 strains. Colour code: green, unique changes that differed from both reference strains; orange, shared residues with the Ingelvac PRRS MLV; blue, shared residues with the wild-type virus strain, VR2332.

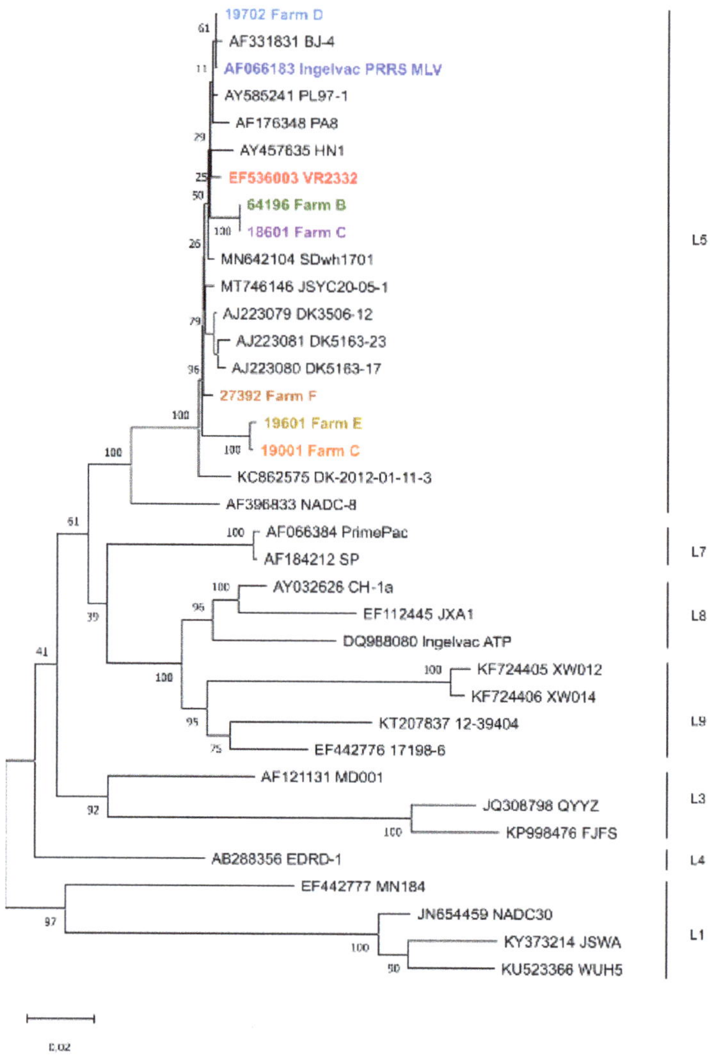

Figure 5. Maximum-likelihood-based phylogenetic tree based on the ORF2-7 region of the genome. The study strains and the reference strains, Lelystad and VR-2332, are highlighted with different colours. For a better resolution of phylogenetic relationships among strains, additional sequences from the GenBank were also used. The lineage specificities are shown on the right.

Phylogenetic analysis was performed to connect genetic data to the epidemiologic observations. The ORF2-7-based phylogenetic tree confirmed the relationship among the Hungarian strains involved in the two PRRSV-2-associated infection chains, and these strains clustered with the Ingelvac PRRS MLV vaccine virus, the parental VR2332 strain and other vaccine-related strains from different countries (Figure 5). Sequences from Farms B and C, likewise C and E, formed two distinct clusters while the sequence originated from Farm D clustered with the Ingelvac PRRS MLV.

4. Discussion

Eradication of PRRSV from the pig population of Hungary is nearing completion. As the number of PRRS-free herds increases, it is a fundamental obligation for the authorities to maintain the PRRS-free status. A key measure in these efforts is the continuous epidemiologic surveillance and the elimination of all PRRS infections, including those resulting from vaccination with live virus. Thus, awareness of the major epidemiological risks greatly contributes to the effective control of PRRS.

The international regulation by the World Organisation for Animal Health (WOAH) requires the verification of the occurrence of PRRS infection according to the following criteria: (i) isolation of PRRSV from pig, except vaccine strains; (ii) detection of PRRSV's antigen or RNA, which is not the consequence of vaccination; (iii) detection of the vaccine PRRSV strain's specific antigen or RNA from non-vaccinated pig; (iv) detection of antibody against PRRSV, which is not the consequence of vaccination (https://www.woah.org/ (accessed on 09 September 2023)). The Hungarian law added a further criterion declaring that fattening units must be free of both wild-type and vaccine-origin PRRSV [6].

From the perspectives of epidemiological judgment and regulation, the vaccine-origin PRRSV strains are considered less harmful, and their impact on economic losses is negligible when compared to infection with wild-type variants. Apparently, this is in accordance with the requirements of vaccine use. However, vaccine viruses introduced without vaccination into a non-vaccinated pig population may serve as a possible source of infection as can be seen in the case of wild-type strains [21–23].

In this study, the dispersal routes and the transmission modes of the NA-type Ingelvac PRRS MLV vaccine origin strain was sought in Hungary via epidemiological investigation and virus genetic analysis. Our investigations led to the conclusion that imported pig stocks that temporarily stayed at collecting stations in Denmark were the primary source of infection. Although the number of samples whose sequence data were available for analysis was limited, our virus genetic data helped clarify the epizootiology of vaccine-origin PRRS infection in the affected swine herds. Analysis of the ORF2-7 region of all available NA type PRRSV study strains showed close genetic relationship to the Ingelvac PRRS MLV strain. Moreover, the deduced amino acid sequences uncovered greater number of shared residues with the vaccine strain than with the wild-type VR2232 strain.

One of the epidemic cases presented here underlines the responsibility of the importer (Farm A), as without the proper examination of imported pigs, a previously scarce, identified live vaccine strain—the Ingelvac PRRS MLV—was introduced into five Hungarian PRRS-free pig herds. The Ingelvac-PRRS-MLV-related variants detected on Farm B (sample 64196) and one obtained from sows on Farm C (sample 18601) showed 99% nt identity to the vaccine strain and clustered together in the phylogenetic tree. This finding implies that these sequences may represent an earlier infection in the transmission route as the strain barely differed from the original vaccine strain. Furthermore, it confirms the spread of PRRS by the slaughterhouse's vehicle operating between these two pig farms during the period of infection. At Farm C, another sequence (sample 19001) was identified that originated from a piglet; this sample shared 97.6% nucleotide identity with sample 18601 (from a sow), suggesting a greater level of virus divergence during the infection cycle among swine of different ages, a finding that has been reported previously [24]. This latter sequence from Farm C and sample 19601 from Farm E showed 98.3% and 98.2% nt identity to the vaccine virus, respectively. The genetic difference suggested that the circulating PRRSV strains in these herds could have been the most genetically distant viruses from the initially imported vaccine virus. A recent paper reported that sequences gathered during outbreaks could accumulate high number of mutations [24]. In Denmark, the Ingelvac PRRS MLV derivative sequences showed a divergence of up to 6% [12], whereas in the USA, less than 5% divergence to the vaccine was observed in most vaccine-related strains [25,26]. Additionally, our previous study also supports the possibility of this level of genetic variation among vaccine-derived PRRSV-2 strains [9]. Results from the phylogenetic analysis were consistent with the theory that PRRSV was introduced to Farm E by the settled fatteners from Farm C. Sample 19702 from Farm D shared 99.9% nucleotide identity with the Ingelvac PRRS MLV strain. We hypothesize that at the very beginning of the vaccine virus transmission, infected fatteners from Farm B were transported to the slaughterhouse. Afterwards, the virus spread to Farm D when the slaughterhouse's vehicle was loaded for weighing of fatteners of Farm D. In this case, this particular virus variant showed less variation compared to the original vaccine virus suggesting that it may have undergone less animal-to-animal passage in different herds. Collectively, tracking the routes of the transmission chain we revealed that within six months (from November 2020 to April 2021), a newly introduced vaccine-origin PRRSV strain was able to spread to several different establishments (in this case four) and could leap great geographic distances that may jeopardize the PRRS-free status of herds in parts of the country.

There are two main factors contributing to the spread of PRRSV-2 in Hungary, based on our previous and current observations: the initial introduction by importing infected pigs and subsequent circulation among different herds [9]. Of note, due to the strict measures implemented in this country, newly introduced strains are eradicated rapidly. However, legal measures to control PRRSV could be hindered if, for example, external biosecurity measures are occasionally neglected in farm management or in case of shared equipment and vehicles or inadequate location of facilities.

Epidemiological investigation convincingly identified the causative virus beyond the accumulation of cases and provided relevant clues concerning the vaccine virus transmission; however, some uncertainties remained. Throughout the study period the detected vaccine-origin PRRSV-2 sequences showed 0.1% to 1.8% nt distance from the Ingelvac PRRS MLV along the ORF2-7 region. The diversity among all strains identified in this study was even greater and reached up to 2.5% nt divergence with some genomic region being even more divergent (ORF5, up to 3.6%). Thus, the observed sequence diversity in the sample set was somewhat greater than expected within this relatively short period of time. However, this is not a unique finding, given that the maximum genome-level sequence divergence detected so far within a year was as high as 6.4% [12]. We speculated that deep sequencing may help define whether a mixture of closely related vaccine strain-origin variants co-circulate because of the viral quasispecies features. From this viewpoint, a limitation of the study was that sampling for sequencing was randomly carried out, and the population structure of vaccine virus derivates could not be determined in simultaneously sampled animals. Despite this limitation, the strategy of choosing deep sequencing uncovered some details concerning the intra-host evolution of vaccine strain from independent herds. We chose a 10% cut-off value to define a minor variant in the sample as a possible marker for the quasispecies structure. Four out of six samples showed evidence of intra-host viral diversity, whereas two samples were genetically homogenous. In samples with multiple co-evolving virus, the SNV sites ranged from 2 to 57 along the 3188 bp fragment coding for the major virion antigens. Seeing this divergence and the ratio of the minor and the major sequence variants, we can conclude that the SNV structure cannot be directly related to the possible mutation rates derived from the chemistry of PCR or the sequencing technology. Indeed, the data obtained in this study, show that the Ingelvac PRRS MLV vaccine strain may become fairly diverse, raising questions concerning the rate of evolution and selection of dominant variants in individual animals as well as in affected herds. An important finding was that the distribution of SNVs did not show a tendency to revert to the exact form of the wild-type parental strain, VR2332, in any samples including those identified in pigs showing clinical manifestations. Different virus genomic evolution was also reported before when analysing the population structure of a single Ingelvac-PRRS-MLV-related strain [27]. We found no correlation between the number of SNVs and the sampling location within the transmission chain either; therefore, it is not clear whether the number of minor genomic variants increases over time in pigs reared in successive facilities. A more thoughtful sampling process in similar situations might help understand if the observed sequence divergence among samples is driven by random effects or if there are yet unidentified driving forces that act on the microevolution of PRRSV genomes.

5. Conclusions

In this paper, two distinct transmission chains of PRRSV-2-associated cases are reported, where epidemiologic investigation highlighted the risks associated with free movement of livestock in the EU. Although PRRS is still a major source of economic losses for the swine industry, it is not a notifiable disease in EU countries and no legislation was implemented for the regulation of pig trade on an EU standard. As reported previously, it is proven that imported prefatteners are a major threat to maintain the PRRS-free status of herds [18]. The ratification of swine stock import solely from PRRS-free pig farms restricted the entry of severely infected pigs to Hungary. Subsequently, an additional critical problem became the focus of attention. The export practice, such as the utilization of collecting stations, is considered as a primary animal health risk for Hungary, particularly in case of PRRSV-2 strains that are not endemic in Hungary. Temporary cohabitation of pigs from different herds at the same facility could be a main source of infection, as the pigs are in different health status and may be in different stages of infection. To minimize the risk of PRRSV infection through import and thereby reduce further damage to farmers, it is recommended that pigs are transported directly from the exporting holdings without the involvement of transit stations. Alternatively, these stations could be designed so that pigs from different herds do not come into contact and thus are unable to transmit infections to each other. In parallel, another critical task to reduce the risk of PRRSV spread is to improve biosecurity and regulatory measures within pig holdings in Hungary.

Author Contributions: Conceptualisation: Á.B., I.N. and I.S.; Data and methodology: S.J., K.B. (Krisztina Bali), I.S. and Á.B.; Validation: I.N. and K.B. (Krisztián Bányai); Formal analysis: I.S., S.J. and Á.B.; Data curation: Á.B., S.J., K.B. (Krisztina Bali) and I.S.; Writing—original draft preparation: S.J., K.B. (Krisztián Bányai) and I.S.; Funding acquisition: Á.B., K.B. (Krisztián Bányai) and I.S. All authors have read and agreed to the published version of the manuscript.

Funding: This work was supported by the National Laboratory for Infectious Animal Diseases, Antimicrobial Resistance, Veterinary Public Health and Food Chain Safety, RRF- 2.3.1-21-2022-00001. This work was also prepared with the professional support of the Doctoral Student Scholarship Program of the Co-operative Doctoral Program of the Ministry of Innovation and Technology financed from the National Research, Development and Innovation Fund.

Institutional Review Board Statement: Not applicable. None of the tests described in the manuscript required the Ethics Committee or Institutional Review Board approval. There were no experimental animals involved in the study.

Informed Consent Statement: Not applicable.

Data Availability Statement: Sequence data were deposited in GenBank (OR143097-OR143102). Additional data are available from Dr. István Szabó via the e-mail address shown on the title page.

Acknowledgments: The authors thank Tamás Molnár, who passed away due to COVID-19 on 20 November 2020, for his devotion to the Hungarian PRRS eradication program.

Conflicts of Interest: The authors declare no conflict of interest. The funding bodies had no role in the design of the study, in the collection, analysis, and interpretation of data or in writing of the manuscript.

References

1. Maclachlan, N.J.; Dubovi, E.J. *Fenner's Veterinary Virology*, 4th ed.; Academic Press/Elsevier: London, UK, 2011; pp. 415–424.
2. Neumann, E.J.; Kliebenstein, J.B.; Johnson, C.D.; Mabry, J.W.; Bush, E.J.; Seitzinger, A.H.; Green, A.L.; Zimmerman, J.J. Assessment of the economic impact of porcine reproductive and respiratory syndrome on swine production in the United States. *J. Am. Vet. Med. Assoc.* **2005**, *227*, 385–392. [CrossRef] [PubMed]
3. Holtkamp, D.J.; Kliebenstein, J.B.; Neumann, E.J.; Zimmerman, J.J.; Rotto, H.F.; Yoder, T.K.; Wang, C.; Yeske, P.E.; Mowrer, C.L.; Haley, C.A. Assessment of the economic impact of porcine reproductive and respiratory syndrome virus on United States pork producers. *J. Swine Health Prod.* **2013**, *21*, 72–84.
4. Snijder, E.J.; Meulenberg, J.J.M. The molecular biology of arteriviruses. *J. Gen. Virol.* **1998**, *79*, 961–979. [CrossRef] [PubMed]

5. Brinton, M.A.; Gulyaeya, A.; Balasuriya, U.B.R.; Dunowska, M.; Faaberg, K.S.; Goldberg, T.; Leung, F.C.-C.; Nauwynck, H.J.; Snijder, E.J.; Stadejek, T.; et al. Expansion of the Rank Structure of the Family Arteriviridae and Renaming Its Taxa. Available online: https://talk.ictvonline.org/taxonomy/p/taxonomy-history?taxnode_id=20186087 (accessed on 9 May 2022).
6. Anonymous. Decree No. 3/2014 (I.16.) VM of the Minister of Rural Development on PRRS eradication of the swine population of Hungary. *Magy. Közlöny* **2014**, *3*, 447. (In Hungarian)
7. Szabó, I.; Bognár, L.; Molnár, T.; Nemes, I.; Bálint, Á. PRRS eradication from swine farms in five regions of Hungary. *Acta Vet. Hung.* **2020**, *68*, 257–262. [CrossRef]
8. Szabó, P.M.; Szalay, D.; Kecskeméti, S.; Molnár, T.; Szabó, I.; Bálint, Á. Investigations on spreading of PRRSV among swine herds by improved minimum spanning network analysis. *Sci. Rep.* **2020**, *10*, 19217. [CrossRef]
9. Jakab, S.; Kaszab, E.; Marton, S.; Bányai, K.; Bálint, Á.; Nemes, I.; Szabó, I. Genetic diversity of imported PRRSV-2 strains, 2005–2020, Hungary. *Front. Vet. Sci.* **2022**, *9*, 1581. [CrossRef]
10. Bøtner, A.; Strandbygaard, B.; Sørensen, K.J.; Have, P.; Madsen, K.G.; Madsen, E.S.; Alexandersen, S. Appearance of Acute PRRS-like Symptoms in Sow Herds after Vaccination with a Modified Live PRRS Vaccine. *Vet. Rec.* **1997**, *141*, 497–499. [CrossRef]
11. Greiser-Wilke, I.; Fiebig, K.; Drexler, C.; Grosse Beilage, E. Genetic Diversity of Porcine Reproductive and Respiratory Syndrome Virus (PRRSV) in Selected Herds in a Pig-Dense Region of North-Western Germany. *Vet. Microbiol.* **2010**, *143*, 213–223. [CrossRef]
12. Kvisgaard, L.K.; Hjulsager, C.K.; Brar, M.S.; Leung, F.C.C.; Larsen, L.E. Genetic Dissection of Complete Genomes of Type 2 PRRS Viruses Isolated in Denmark over a Period of 15 Years. *Vet. Microbiol.* **2013**, *167*, 334–344. [CrossRef] [PubMed]
13. Stadejek, T.; Stankevicius, A.; Murtaugh, M.P.; Oleksiewicz, M.B. Molecular Evolution of PRRSV in Europe: Current State of Play. *Vet. Microbiol.* **2013**, *165*, 21–28. [CrossRef] [PubMed]
14. Indik, S.; Schmoll, F.; Sipos, W.; Klein, D. Genetic Variability of PRRS Virus in Austria: Consequences for Molecular Diagnostics and Viral Quantification. *Vet. Microbiol.* **2005**, *107*, 171–179. [CrossRef] [PubMed]
15. Arruda, A.G.; Tousignant, S.; Sanhueza, J.; Vilalta, C.; Poljak, Z.; Torremorell, M.; Alonso, C.; Corzo, C.A. Aerosol Detection and Transmission of Porcine Reproductive and Respiratory Syndrome Virus (PRRSV): What Is the Evidence, and What Are the Knowledge Gaps? *Viruses* **2019**, *11*, 712. [CrossRef]
16. Noordhuizen, J.P.T.M.; Frankena, K.; Thrusfield, M.V.; Graat, E.A.M. *Application of Quantitative Methods in Veterinary Epidemiology*, 2nd ed.; Wageningen Pers: Wageningen, The Netherlands, 2001; p. 429.
17. Balka, G.; Hornyák, Á.; Bálint, Á.; Kiss, I.; Kecskeméti, S.; Bakonyi, T.; Rusvai, M. Genetic diversity of porcine reproductive and respiratory syndrome virus strains circulating in Hungarian swine herds. *Vet. Microbiol.* **2008**, *127*, 128–135. [CrossRef] [PubMed]
18. Szabó, I.; Molnár, T.; Nemes, I.; Abonyi, T.; Terjék, Z.; Bálint, Á. PRRSV eradication on large-scale fattening pig farms in Hungary between 2014 and 2019. *Acta Vet. Hung.* **2019**, *67*, 529–542. [CrossRef]
19. Bali, K.; Bálint, Á.; Farsang, A.; Marton, S.; Nagy, B.; Kaszab, E.; Belák, S.; Palya, V.; Bányai, K. Recombination events shape the genomic evolution of infectious bronchitis virus in Europe. *Viruses* **2021**, *13*, 535. [CrossRef]
20. Kumar, S.; Stecher, G.; Li, M.; Knyaz, C.; Tamura, K. MEGA X: Molecular Evolutionary Genetics Analysis across computing platforms. *Mol. Biol. Evol.* **2018**, *35*, 1547–1549. [CrossRef]
21. Storgaard, T.; Oleksiewicz, M.; Bøtner, A. Examination of the selective pressures on a live PRRS vaccine virus. *Arch. Virol.* **1999**, *144*, 2389–2401. [CrossRef]
22. Chen, N.; Li, X.; Xiao, Y.; Li, S.; Zhu, J. Characterization of four types of MLV-derived porcine reproductive and respiratory syndrome viruses isolated in unvaccinated pigs from 2016 to 2020. *Res. Vet. Sci.* **2021**, *134*, 102–111. [CrossRef]
23. Bálint, Á.; Molnár, T.; Kecskeméti, S.; Kulcsár, G.; Soós, T.; Szabó, P.M.; Kaszab, E.; Fornyos, K.; Zádori, Z.; Bányai, K.; et al. Genetic variability of PRRSV vaccine strains used in the National Eradication Programme, Hungary. *Vaccines* **2021**, *9*, 849. [CrossRef]
24. Clilverd, H.; Martín-Valls, G.; Li, Y.; Martín, M.; Cortey, M.; Mateu, E. Infection dynamics, transmission, and evolution after an outbreak of porcine reproductive and respiratory syndrome virus. *Front. Microbiol.* **2023**, *14*, 1109881. [CrossRef] [PubMed]
25. Kikuti, M.; Sanhueza, J.; Vilalta, C.; Paploski, I.A.D.; VanderWaal, K.; Corzo, C.A. Porcine reproductive and respiratory syndrome virus 2 (PRRSV-2) genetic diversity and occurrence of wild type and vaccine-like strains in the United States swine industry. *PLoS ONE* **2021**, *16*, e0259531. [CrossRef] [PubMed]
26. Rupasinghe, R.; Lee, K.; Liu, X.; Gauger, P.C.; Zhang, J.; Martínez-López, B. Molecular evolution of Porcine Reproductive and Respiratory Syndrome Virus field strains from two swine production systems in the midwestern United States from 2001 to 2020. *Microbiol. Spectr.* **2022**, *10*, e0263421. [CrossRef] [PubMed]
27. Brar, M.S.; Shi, M.; Hui, R.K.H.; Leung, F.C.C. Genomic evolution of Porcine Reproductive and Respiratory Syndrome Virus (PRRSV) isolates revealed by deep sequencing. *PLoS ONE* **2014**, *9*, e88807. [CrossRef]

animals

MDPI

Article

Co-Circulation of Multiple Coronavirus Genera and Subgenera during an Epizootic of Lethal Respiratory Disease in Newborn Alpacas (Vicugna pacos) in Peru: First Report of Bat-like Coronaviruses in Alpacas

Luis Llanco [1], Karubya Retamozo [2], Noriko Oviedo [2], Alberto Manchego [2], César Lázaro [3], Dennis A. Navarro-Mamani [4], Norma Santos [5] and Miguel Rojas [2,*]

1 Escuela de Medicina Humana, Universidad Privada San Juan Bautista, Apartado, Chincha 15067, Peru; luis.llanco@uspsjb.edu.pe
2 Laboratório de Inmunología, Facultad de Medicina Veterinaria, Universidad Nacional Mayor de San Marcos, Apartado, Lima 03-5137, Peru; karu.3.93@gmail.com (K.R.); yahaira.oviedo@unmsm.edu.pe (N.O.); amanchegos@unmsm.edu.pe (A.M.)
3 Laboratório de Farmacología y Toxicología Veterinaria, Facultad de Medicina Veterinaria, Universidad Nacional Mayor de San Marcos, Apartado, Lima 03-5137, Peru; clazarod@unmsm.edu.pe
4 Laboratório de Virología, Facultad de Medicina Veterinaria, Universidad Nacional Mayor de San Marcos, Apartado, Lima 03-5137, Peru; dnavarrom@unmsm.edu.pe
5 Instituto de Microbiologia Paulo de Góes, Universidade Federal do Rio de Janeiro, Rio de Janeiro 21941-902, RJ, Brazil; nsantos@micro.ufrj.br
* Correspondence: mrojasm2@unmsm.edu.pe; Tel.: +51-98687439

Simple Summary: Alpacas (Vicugna pacos) and llamas (*Lama glama*) constitute the most significant livestock wealth of the Andean populations of South America. Infectious diseases, particularly respiratory and enteric infections, cause high morbidity and mortality in offspring and adult animals. In this study, we demonstrated that multiple variants of the coronavirus co-circulated among Peruvian alpacas. We also demonstrated that some of these variants bear similarities to coronavirus strains detected in bats. For a better understanding of the infections that afflict these animals, continuous surveillance is needed to identify the emergence of new genotypes and viral variants that are potentially pathogenic to alpacas and humans. Future studies should include the sequencing of genes encoding CoV spike proteins and host receptors to confirm interspecies transmission.

Abstract: Coronaviruses (CoVs) infect a wide range of hosts, including humans, domestic animals, and wildlife, typically causing mild-to-severe respiratory or enteric disease. The main objective of this study was to identify CoV genera and subgenera detected in Peruvian alpacas. Lung lavage specimens were collected from 32 animals aged 1 to 6 weeks. CoVs were identified by using RT-PCR to amplify a pan-CoV conserved region of the RNA-dependent RNA polymerase-encoding gene. A nested PCR was performed to identify β-CoVs. Then, β-CoV-positive samples were subjected to RT-PCR using specific primers to identify the *Embecovirus* subgenus. Out of 32 analyzed samples, 30 (93.8%) tested positive for at least one CoV genus. β-, α-, or unclassified CoVs were identified in 24 (80%), 1 (3.3%), and 1 (3.3%) of the positive samples, respectively. A CoV genus could not be identified in two (6.7%) samples. A mixture of different CoV genera was detected in two (6.7%) samples: one was co-infected with β- and α-CoVs, and the other contained a β- and an unclassified CoV. A sequence analysis of the amplicons generated by the PCR identified 17 β-CoV strains belonging to the subgenus *Embecovirus* and two α-CoV strains belonging to *Decacovirus*. A phylogenetic analysis of two strains revealed a relationship with an unclassified *Megaderma* BatCoV strain. A subgenus could not be identified in nine β-CoV samples. Our data show a high prevalence and a high genetic diversity of CoV genera and subgenera that infect alpacas, in which the β-CoV subgenus *Embecovirus* predominated. Our data also suggest a new role for bats in the dissemination and transmission of uncommon CoVs to alpacas raised in rural Peru.

Keywords: alpaca; BatCoV; β-CoV; Embecovirus; α-CoV; decacovirus; mortality; Peru

1. Introduction

Raising alpacas (Vicugna pacos) and llamas (*Lama glama*) is the main economic activity of the Andean populations of southern Peru [1]. The Peruvian alpaca herd represents 85% of the world's population of these animals. Breeding is distributed primarily (77%) in the southern highlands in the departments of Puno and Cuzco [2–4]. Most alpaca farms are small (50–100 animals) and conduct extensive breeding with inappropriate livestock management [2,3,5].

Neonatal mortality in alpacas can reach up to 30% [2], which is primarily due to respiratory and enteric infections following the failure of passive transfer colostrum antibodies [6,7]. Co-circulating viruses and bacteria have been implicated in outbreaks that feature co-infections due to highly lethal pathogens such as *Streptococcus pneumoniae, Mannheimia haemolytica,* and *Pasteurella multocida* [8]; coronavirus (CoV), mammalian orthoreovirus, and rotavirus *A* [9–12]; CoV and *Salmonella* spp. [13]; parainfluenza virus type 3, bovine respiratory syncytial virus, *Pasteurella multocida,* and *Mannheimia haemolytica* [14,15].

CoVs infect a wide range of hosts, including humans, domestic animals, and wildlife, typically causing mild-to-severe respiratory or enteric disease [16]. This family of viruses, with a positive-sense single-stranded RNA genome, exhibits a high genetic diversity and is classified into four genera: *Alphacoronavirus* (α-CoV), *Betacoronavirus* (β-CoV), *Gammacoronavirus* (γ-CoV), and *Deltacoronavirus* (δ-CoV) [17]. In general, α-CoV and β-CoV infect mammalian hosts, while γ-CoV and δ-CoV infect birds, although some of them also infect mammals. It has been suggested that bats are the genetic source of α-CoV and β-CoV, while birds are the genetic source of γ-CoV and δ-CoV. Bat CoVs, in addition to infecting several bat species, cross the interspecies barrier infecting other mammals, including humans. Similarly, CoVs from birds have acquired the ability to infect a variety of bird species and, occasionally, some mammalian species, such as whales and pigs [18]. Each genus is further divided into subgenera and species, characterized by great genetic diversity resulting from the high frequency of homologous recombination and the accumulation of point mutations, which confer the ability to cross species barriers [17,19]. Recurrent events of interspecies transmission represent the potential for accelerating viral evolution and, consequently, the possibility of the emergence of new viral strains. The cohabitation of birds and mammals in domestic and wild environments, as well as proximity to humans, may offer the possibility of crossing the interspecies barrier and eventually lead to the emergence of new variants capable of adapting to new hosts, including humans, as observed in SARS-CoV-1, SARS-CoV-2, and MERS-CoV [17,18,20]. Based on phylogenetic analyses, it appears that all human coronaviruses have animal origins: SARS-CoV, MERS-CoV, HCoV-NL63, and HCoV-229E are considered to have originated in bats; HCoV-OC43 and HKU1 likely originated in rodents [17]. Domestic animals may have played important roles as intermediate hosts that allow for the transmission of the virus from natural hosts to humans. The camelids were likely intermediate hosts of HCoV-229E [21,22], and HCoV-OC43 likely evolved from ancestral BCoV strains that crossed the interspecies barrier and established infections in humans [23,24].

SARSCoV-1, SARSCoV-2, and MERS-CoV are examples of viruses that emerged in the human population after spillover events, likely from an animal reservoir, with devastating effects on public health; these viruses are classified as β-CoVs that originated from bats, which are transmitted to humans through intermediate hosts such as civets, pangolins, and old-world camelids, respectively [17,20]. Natural and experimental infections by SARS-CoV-2 have been described in a wide variety of animal hosts. Ferrets and cats were found to be highly susceptible to the virus, while dogs are less susceptible, and chickens, ducks, and pigs have shown lower susceptibility [25,26]. The free-ranging white-tailed deer has also been shown to be highly susceptible to SARS-CoV-2 virus infection and capable of sustaining transmission in nature [27]. Serological evidence has demonstrated the ability of SARS-CoV-2 to naturally infect small ruminants such as cattle, sheep, goats, and dromedary camels [28,29]. A SARS-CoV-2 spillback transmission from humans to animals has been suggested [30–32], raising the possibility of SARS-CoV-2 amphixenosis.

Bovine CoVs (BCoVs) that replicate in the intestine, infect the upper and lower respiratory systems, and are commonly associated with enteric disease in cattle. Domestic (goats, sheep, water buffalos, dromedary camels, alpacas, and llamas) and wild ruminants (reindeer, elk, sambar deer, sika deer, musk oxen, wisents, wood bison, waterbucks, sitatungas, stable antelopes, nyalas, giraffes, and Himalayan tahrs) are infected by CoV strains that share biological, antigenic, and genetic similarities with BCoVs (called bovine-like CoVs or BCoV-likes) [16,33]. BCoV-likes have also been detected in other species such as Indonesian tapirs (*Acrocodia indica*), an ungulate but non-ruminant species, with dysentery [34]; dogs with respiratory disease [35]; and humans with diarrhea [36], showing the ability of BCoV-likes to adapt to new hosts.

CoVs detected in Peruvian and other South American alpacas have been associated with enteritis caused by bovine-like coronavirus strains [10,11,37,38], identified as a β-CoV of the *Embecovirus* subgenus. In North America, viruses belonging to the β-CoV genus (subgenus *Embecovirus*) and α-CoV (subgenus *Duvinacovirus*) have been reported in alpacas with enteric and respiratory diseases, respectively [13,39,40].

This study aimed to identify the genera and subgenera of CoVs present in bronchial lavage samples obtained from newborn alpacas in Cuzco.

2. Materials and Methods

2.1. Sampling

Lung lavage specimens were collected from newborn alpacas (n = 32) between one and six weeks of age from the rural community of Silly, located in the District of Marangani, Province of Canchis, Department of Cuzco, Peru (14°21′12″ S, 71°10′17″ W, 3800 masl) during the birthing season of 2012. Samples were collected directly from lungs during necropsy and stored at −70 °C until processing at the Laboratory of Veterinary Virology and Immunology at the Facultad de Medicina Veterinaria of the Universidad Nacional Mayor de San Marcos (FMV-UNMSM), Lima, Peru.

2.2. Viral Detection and Identification

Viral RNA was extracted from the lung lavage using TRIzol™ LS Reagent (Thermo Fisher Scientific, Waltham, MA, USA) according to the manufacturer's instructions. Samples were tested for the presence of CoVs via reverse-transcription PCR amplification (RT-PCR) and nested PCR using specific primers targeting a 251 bp fragment of the RNA-dependent RNA polymerase gene (RdRp), which is conserved across all CoVs (Table 1). Briefly, the viral RNA was subjected to one reverse-transcription cycle consisting of 5 min at 25 °C followed by 45 min at 42 °C and one step of 2 min at 95 °C followed by PCR cycles as described elsewhere [41]. The generated amplicons were submitted to nested PCR using specific primers for β-CoV to generate a 227 bp from RpRd [9]. β-CoV-positive samples were further analyzed to identify subgenus *Embecovirus* using nested PCR. The PCR conditions were as previously described by Brandão et al., 2004 [42] (Table 1). PCR products were separated with 1.5% (*w*/*v*) agarose gel electrophoresis, stained with ethidium bromide, and visualized under UV light. A 100 bp DNA ladder (Promega, Madison, WI, USA) was used to determine molecular size.

To validate the PCR assays, positive controls were used for each of the four CoV genera (α-CoV, β-CoV, γ-CoV, and δ-CoV), which included CoVs isolated from pigs, chickens, and alpacas and belonging to the collection of the Laboratory of Veterinary Virology and Immunology of the FMV-UNMSM. The alpaca CoV strains used as positive controls were AlpCoV-SA44 and AlpCoV-HN (GenBank accession numbers KX266949 and KX266944, respectively), both belonging to β-CoV; subgenus, *Embecovirus*; species, bovine-like CoV.

Table 1. Primers used in the RT-PCR and nested PCR assay for CoV detection †.

Virus	Gene	Assay	Primer *	Primer Sequence 5′ → 3′	Position	Product Size (bp)	Reference
All CoV	RdRd +	RT-PCR	Cor-FW	ACWCARHTVAAYYTNAARTAYGC	14,922–14,944	251	[41]
			Cor-RV	TCRCAYTTDGGRTARTCCCA	15,153–15,172		
β-CoV	RdRd	Nested PCR	Beta.CoV.F	ATTAGTGCWAAGAATAGAGCYCGCAC	14,946–14,971	227	[9]
			Beta.CoV.R	TCACAYTTWGGRTARTCCCADCCCA	15,148–15,172		
Embecovirus	RdRd	Nested PCR	CV2U.F	TACTATGACTGGCAGAATGTTTCA	14,996–15,019	136	[42]
			CV2L.R	AACATCTTTAATAAGGCGRCGTAA	15,108–15,131		

† CoV = coronavirus; + RdRp = RNA-dependent RNA polymerase. * The primers' positions were determined based on the reference CoV strain DQ915164.

2.3. CoV Characterization via Phylogenetic Analysis of Partial Sequences of the RdRp Gene

Amplified genomic segments were sequenced by Macrogen Inc. (Seoul, Republic of Korea). Overlapping sequences were assembled and edited using SeqMan, EditSeq, and MegAlign in the Lasergene software package (Version 7.0, DNASTAR, Madison, WI, USA). Phylogenetic analysis was performed with the MEGAX software [43,44]. Dendrograms were constructed using the maximum likelihood method based on the Hasegawa–Kishino–Yano model [43]. Statistical significance was estimated via bootstrap analysis with 1000 pseudoreplicates. Sequences were compared with reference CoV strains obtained from GenBank (https://www.ncbi.nlm.nih.gov/nucleotide/, accessed on 1 August 2023). Sequences generated in this study were deposited in GenBank under accession numbers OQ845932–OQ845939.

3. Results

3.1. Detection and Identification of CoV Genera and Subgenera

Of the 32 lavage samples, 30 (93.8%) tested positive for at least one CoV genus. Twenty-six samples were positive for β- (24; 80%), α- (1; 3.3%), or unclassified (1; 3.3%) CoVs. In two (6.7%) RT-PCR-positive samples, the CoV genus could not be identified with either nested PCR or sequencing (Figure 1; Table 2). The detection of multiple CoV genera was observed in two (6.7%) samples. One sample contained a mixture of β- and α-CoVs, and the other contained β-CoV and an unclassified CoV (Table 2). The identification of the subgenera of the detected strains showed that 17 strains of β-CoV belonged to the subgenus *Embecovirus*; two α-CoV strains belonged to the genus *Decacovirus*. Two strains showed phylogenetic relationships to an unclassified *Megaderma* bat-CoV strain. Subgenera could not be identified for nine β-CoV strains (Figure 2; Table 2).

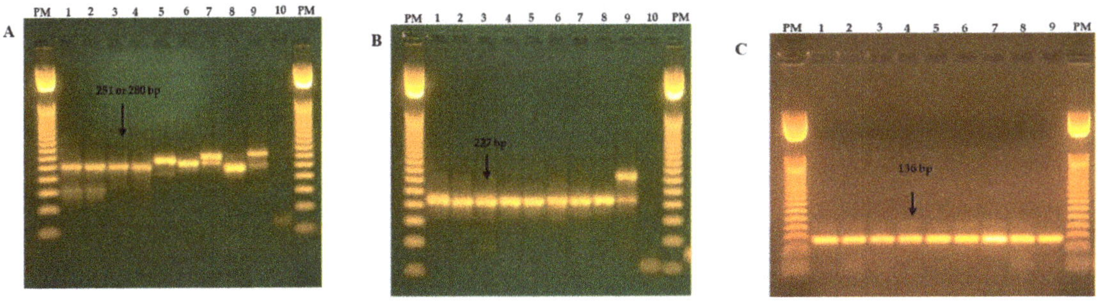

Figure 1. Ethidium bromide-stained 1.5% agarose gel showing the amplified products of a pancoronavirus RT-PCR (251 or 280 bp) from a nested-PCR of betacoronavirus (227 bp) and an Embecovirus nested-PCR (136 bp). (**A**) RT-PCR pancoronavirus (CoV, RdRp gene): PM line, 50 bp of DNA marker; lines from 1 to 9; lung lavage samples; lane 10, negative control. (**B**) Nested-PCR (β-CoV, RdRp gene): PM line, 50 bp of DNA marker; lines from 1 to 9; lung lavage samples; lane 10, negative control. (**C**) Nested-PCR (Embecovirus, RdRp gene): PM line, 50 bp of marker DNA; lines from 1 to 9; lung lavage samples. Note: Pancoronavirus products in alpacas can vary in size from 251 to 280 bp; in some cases, both bands are present; PM = molecular weight.

Table 2. Identification of genera and subgenera of coronaviruses in samples from lung lavages of Peruvian alpacas killed in a respiratory outbreak.

Infection Type	Genus	Subgenus	N° Positive Samples
Single detection	*Betacoronavirus* (*β-CoV*)	*Embecovirus* (EmbeCoV)	15
		Not identified	9
	Alphacoronavirus (*α-CoV*)	*Decacovirus* (DecaCoV)	1
	Unclassified (Megaderma Bat-CoV-like)	Unclassified (Megaderma Bat-CoV-like)	1
	Not identified	Not identified	2
Multiple detection	*β-CoV* + *α-CoV*	EmbeCoV + DecaCoV	1
	β-CoV + Unclassified (Megaderma Bat-CoV-like)	EmbeCoV + unclassified	1
Total			30

β-CoV = *Betacoronavirus*; EmbeCoV = *Embecovirus*; *α-CoV* = *Alphacoronavirus*; DecaCoV= *Decacovirus*; Unclassified = CoV without a determined genus (ICTV, 2023); not identified = products of 251 bp without success in sequencing.

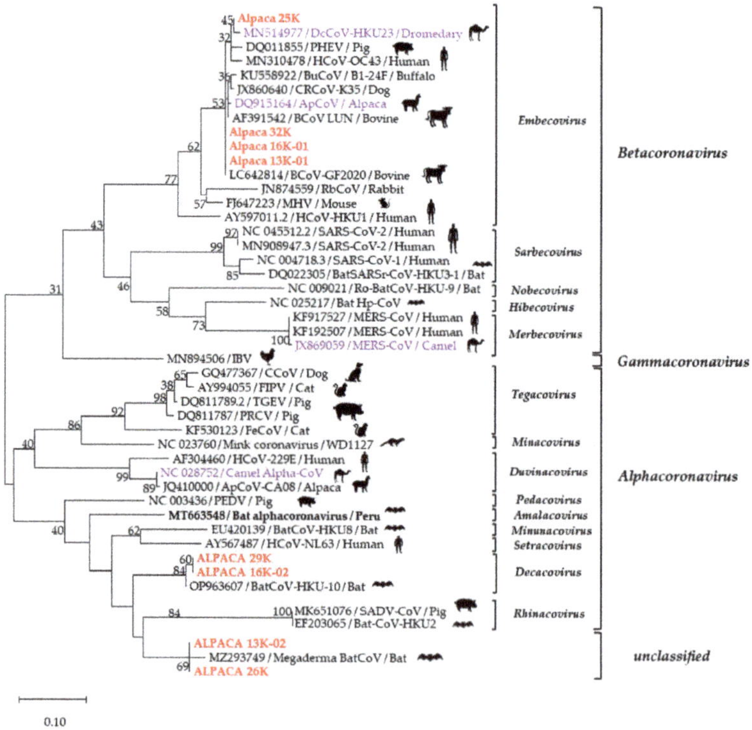

Figure 2. Phylogenetic relationships of coronaviruses based on a 251 bp fragment of RdRp gene. Sequences in red are of our study, and sequences in purple are of different CoV species reported in other camelids. This analysis involved 46 nucleotide sequences identified with GenBank accession numbers. Dendrograms were constructed using the maximum likelihood method; the distances were corrected with the Hasegawa–Kishino–Yano model. Statistical support was provided by bootstrapping 1000 pseudoreplicates. Bootstrap values >75% are given at branch nodes. The distance scale reflects substitutions/sites.

3.2. Phylogenetic Analysis of the RdRp Gene

Of the eighteen products selected for sequencing the partial *RdRp* gene (251 bp), eight were successfully sequenced: 251 bp of four samples (Alp 13K-01, Alp 16K-01, Alp 25K, and Alp 32K) and 112 bp of four other samples (Alp 13K-02, Alp 16K-02, Alp 26K, and Alp 29K). A phylogenetic analysis grouped the sequences into three distinct clusters, each belonging to a different genus: β-CoV (EmbeCoV) or α-CoV (DecaCoV). Two sequences were grouped with a bat coronavirus (BatCoV) strain not yet classified by the ICTV (Figure 2).

Strains 13K-01, 16K-01, 25K, and 32K shared a 98.8%-to-100% nucleotide identity. When compared with CoV reference sequences, they showed a phylogenetic relationship closer to strains of the subgenus *Embecovirus,* with nucleotide identities in a range of 80.9% to 100%, phylogenetically closer to CoVs detected in bovines (BCoV) (LC642814/GF2020 and AF391542 /LUN), alpaca (ApCoV) (DQ915164), and dromedary (DcCoV) (MN514977), with nucleotide identities of 98.8–100%, 98.4–99.6%, 99.2–98%, and 99.2–98.8%, respectively. On the other hand, these strains showed 65.4%, 69.9%, 63.2%, and 58.1% nucleotide identities with reference strains of the β-CoV subgenera *Sarbecovirus, Nobecovirus, Hibecovirus,* and *Merbecovirus,* respectively.

The amplicon positives yielded by pancoronavirus and panbetacoronavirus RT-PCRs that were negative for the *Embecovirus* subgenus could not be sequenced correctly with the Sanger method because they presented excessive noise during chromatography.

Sequences 29K and 16K-02 were 100% identical and showed a phylogenetic relationship to the BatCoV/HKU strain (OP963607; 99.1% nucleotide identity), which is classified as an α-CoV (DecaCoV). Sequences 13K-02 and 26K were identical and phylogenetically closest (98.1% nucleotide identity) to the *Megaderma* CoV strain (MZ293749/Bat-CoV), which is not yet classified in any coronavirus genus. The phylogenetic distance obtained by aligning the 29K strain with the 32K, 25K, and 26K strains resulted in nucleotide identities of 66.9%, 64.6%, and 86.5%, respectively. Co-infections were confirmed by sequencing two samples: 16K (β-CoV + α-CoV) and 13K (β-CoV + unclassified CoV) (Figure 2, Table 2).

4. Discussion

Our results demonstrated the presence of CoV in 93.8% (32/30) of lung lavage samples from newborn alpacas from Cuzco, Peru. Other reports have described outbreaks of diarrhea in alpacas from rural communities located in the Departments of Cuzco and Junín, presenting rates of 87.5% (70/80) and 53.3% (32/60), respectively [9,45].

The β-CoV genus was the most prevalent (86.7%; 26/30), consistent with the data presented by Castilla et al. [9], which detected β-CoVs associated with diarrhea in neonatal alpacas (94.3%; 66/70). However, in that outbreak, the *Embecovirus* subgenus was identified in only 22.9% (16/70) of samples. Furthermore, the subgenus of 71.4% (50/70) of the β-CoV strains could not be identified [9]. On the other hand, in our study, Embecoviruses represented the majority (65%; 17/26) of strains, and yet, subgenus identification was not possible for 35% (9/26) of the strains. These data reveal the co-circulation of distinct viral variants in the studied region. The variation in the prevalence of Embecoviruses could be multifactorial. Although both studies were conducted in the same geographic area, in the rural community of Silly during the alpaca birthing season (January and February), clinical samples were obtained from different anatomic sites (fecal versus lung lavage specimens), which could suggest differential tissue tropisms. On the other hand, the year during which the samples were obtained differed between studies. Castilla et al. analyzed samples obtained in 2015, while those analyzed in this study were collected in 2012. Given the CoV capacity for rapid evolution [17], β-CoV strains belonging to one or more subgenera, distinct from *Embecovirus,* that were already circulating in alpaca herds in 2012 may have adapted to this host, becoming more prevalent over the years and subsequently representing more than 70% of circulating strains in 2015. Therefore, it is essential to monitor these viruses continually to better understand their dynamics in the environment.

In the present study, we report a minor percentage of CoV-positive samples in which a genus could not be identified, similar to previous findings [9]. Our attempts to isolate these viruses in Vero cell cultures (African green monkey kidney cells) were unsuccessful, perhaps because of the long storage period of the samples. These results indicate that studies using next-generation sequencing techniques and continuous molecular surveillance are necessary to better understand the epidemiology and genetic diversity of CoVs in Peruvian alpacas.

The sequencing of four strains (Alp16K-01, 13K-01, 25K, and 32K), classified as β-CoVs/*Embecoviruses*, showed a close relationship with BCoVs, in agreement with previous studies suggesting that either the alpaca and BCoVs arose from a common ancestor or that BCoVs are continuously transmitted to alpacas [11,38,39]. Although South American camelids have been in contact with cattle for about five centuries, recent reports of disease associated with BCoVs may be due to the emergence of new CoV variants that are pathogenic to both bovines and camelids [13,39]. These results can also be explained by geographic and climatologic conditions, where different animal species (pigs, cows, llamas, horses, etc.) coexist and graze in proximity [45].

Two strains (16K-02 and 29K) were classified as α-CoVs/*Decacoviruses* closely related to a bat CoV (bat CoV/HKU; OP963607). This is not the first report of α-CoV infections in alpacas; however, the previously reported subgenus was *Duvinacovirus*, closely related to human coronavirus 229E (HCoV-229E), one of the most prevalent common cold coronaviruses in humans [40]. Subsequent genetic studies suggest that the progenitor of HCoV-229E is an African bat CoV and that camelids were probably the first intermediate hosts that facilitated transmission to humans [17,21,46,47]. This suggests that bats located in the region under study may be transmitting various CoV species with the capacity to adapt and become pathogenic to new host species. To the best of our knowledge, our study is the first to report α-CoV infections in Peruvian alpacas.

A genus was not identified in a pair of samples (13K-02 and 26K); these sequences are genetically very similar to a bat CoV strain (MZ293749/*Megaderma* CoV), which remains unclassified. This confirms both the great diversity of this viral family and the role of bats as an important reservoir of these and other unclassified viruses [48–50].

An important observation in this study was the detection of multiple CoV genera in two samples confirmed through the sequencing of a fragment of *RdRp*. Genomic recombination has been shown to be an important factor in CoV evolution but requires that different viral strains infect the same host and the same cells simultaneously [17,51,52]. In fact, So et al. demonstrated that a strain of MERS-CoV that infects African dromedary camels (DcCoV-HKU23; subgenus, *Embecovirus*) was a recombinant of bovine, rabbit (Rb-CoVHKU14), and rodent CoVs [51]. We suggest that CoVs circulating in alpacas are in active recombination, which favors the evolution of β-, α-, and other CoV genera not yet detected or classified.

Because bats are widely distributed in Peru, they could be the source of the CoV strains in samples 13K-02, 16K-02, 26K, and 29K. For example, *Desmodus rotundus*, a hematophagous bat observed at up to 3680 masl, inhabits the South Andean region (Cusco); alpacas also share this geographic space (Figure 3) [53,54]. The invasion of their habitat, a hematophagous diet, migration due to climate change, and proximity to alpacas may favor interspecies spillover. The transmission of enzootic bat viruses to domestic animals was reported previously. The HKU2-related BatCoV caused a large-scale epizootic of enteric disease in China, resulting in the deaths of more than 24,000 piglets. This outbreak demonstrated that the spillover of a BatCoV can cause severe disease in livestock [17,55]. Bat species are a common origin for most CoVs affecting humans [21]. The intermediate hosts for these CoVs usually belong to other mammalian species [46]. The general ecological separation between bats and humans implies the need for other mammals to act as link hosts between bats and humans [56].

Bergner et al. [54], using metagenomics, characterized viral communities from salivary and fecal specimens obtained from wild bats captured in caves and trees located in high Andean and jungle areas of the Departments of Cusco, Ayacucho, and Amazonas in Peru in a period from 2015 to 2016 (Figure 3) and demonstrated that a subset of CoVs that infect neonatal alpacas—specifically, α-CoV (*Decacovirus* subgenus in our study) and other CoVs of a currently unclassified genus—are products of interspecies transmission between bats and alpacas. This can also be explained because the area where these α-CoVs were identified is endemic to bats [53,54] and close to the collection area of our study.

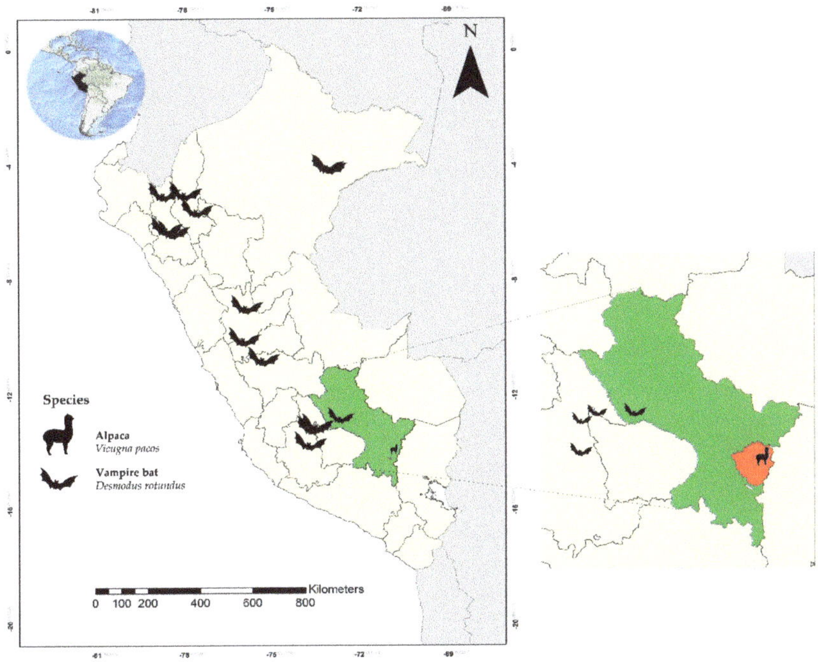

Figure 3. Location of Peruvian bat sampling sites in a study by Bergner [54]. In green, the Department of Cuzco, Peru, and in red, the sampling site of the alpacas for our study.

5. Conclusions

Our data showed the high prevalence and the high genetic diversity of CoV genera and subgenera detected in alpacas, in which the β-CoV subgenus *Embecovirus* predominated. Our data also demonstrated the genetic similarity between strains of CoVs circulating in Peruvian alpacas and bats. Given the high transmissibility and the zoonotic nature of coronaviruses, continuous surveillance is necessary to identify the emergence of new viral genotypes and variants potentially pathogenic to alpacas and humans. Future studies should include the sequencing of genes encoding CoV spike proteins and host receptors to confirm interspecies transmission.

Author Contributions: Conceptualization, M.R., A.M. and N.S.; methodology, L.L., K.R., N.O., M.R., A.M. and D.A.N.-M.; writing—original draft preparation, L.L., M.R. and C.L.; writing—review and editing, M.R., C.L., L.L. and N.S.; funding acquisition, M.R. All authors have read and agreed to the published version of the manuscript.

Funding: This study was funded by the Programa Nacional de Investigación y Estudios Avanzados (PROCIENCIA)–CONCYTEC en el marco de la convocatoria de Proyecto Investigación Básica 2019-01 (grant number 355-2019-FONDECYT), the Vicerrectorado de Investigación y Posgrado de la Universidad Nacional Mayor de San Marcos en el marco de la convocatoria "Proyecto de Investigación con financiamiento para Grupos de Investigación 2021" PCONFIGI–N° A21080551, and the Conselho Nacional de Desenvolvimento Científico e Tecnológico, Brazil (CNPq grant 302275/2022-3).

Institutional Review Board Statement: The procedures and techniques performed on alpacas and llamas used in the present study were approved by the Animal Research Ethics Committee of the Facultad de Medicina Veterinaria, Universidad Nacional Mayor de San Marcos (FMV-UNMSM) and the Instituto Veterinario de Investigaciones Tropicales y de Altura (IVITA), protocol No. 2009-0001. Samples were taken following animal-handling protocols authorized by the Ethics and Animal Welfare Committee of the FMV-UNMSM.

Informed Consent Statement: Not applicable.

Data Availability Statement: Not applicable.

Acknowledgments: We thank Miguel Angel Villareal Balboa for technical assistance and Elena Yudith Carrasco Charquero for being the administrative coordinator of "Proyecto Investigación Básica 2019-01" (grant number 355-2019). This work is dedicated to the memory of **Lina Mercedes Montes Collave**; thanks for everything, Mother.

Conflicts of Interest: The authors declare no conflict of interest. None of the authors of this paper has a financial or personal relationship with other people or organizations that could inappropriately influence the content of the paper. The funders had no role in the design of the study; in the collection, analyses, or interpretation of the data; in the writing of the manuscript; or in the decision to publish the results.

References

1. Méndez, M.P. Herencia textil, identidad indigena y recursos económicos en la Patagonia Argentina. Estudio de un caso: La comarca de La Meseta Central de La Provincia de Chubut. *AIBR-Rev. Antropol. Iberoam.* **2009**, *4*, 11–53. [CrossRef]
2. Consejo Nacional de Camélidos Sudamericanos (CONACS). Estrategia Nacional de Desarrollo: Los Camélidos Domésticos en el Perú. *Ministerio de Agricultura.* Available online: https://www.whyalpaca.com/assets/documentos/estrategia_nacional.pdf (accessed on 31 July 2023).
3. FAO (Food and Agriculture Organization of the United Nations). Situación Actual de los Camelidos Sudamericanos en el Perú. *Organización de las Naciones Unidas Para la Agricultura y Alimentación. Proyecto de Cooperación Técnica en Apoyo a la Crianza y Aprovechamiento de los Camelidos Sudamericanos en la Region Andino.* Available online: http://tarwi.lamolina.edu.pe/~emellisho/zootecnia_archivos/situacion%20alpcas%20peru.pdf (accessed on 31 July 2023).
4. MINAGRI (Ministerio de Desarrollo Agrario y Riego/Peru). Situacion de las Actividades de Crianza y Producción de Camelidos Sudamericanos en el Peru. MINAGRI. Available online: https://www.midagri.gob.pe/portal/datero/40-sector-agrario/situacion-de-las-actividades-de-crianza-y-producci/298-camelidos-sudamericanos?start=2 (accessed on 31 July 2023).
5. Ramon, P.T. *Trayectorias Comunales Cambios y Continuidades en Comunidades Campesinas e Indígenas del sur Andino, Segunda*; GRUPO Propussta Ciudadana, Ed.; Fordfoundation: Lima, Perú, 2019.
6. Pinn, T.L.; Gagliardo, L.F.; Purdy, S.R.; Appleton, J.A.; Stokol, T. Comparison of three immunoglobulin G assays for the diagnosis of failure of passive transfer of immunity in neonatal alpacas. *J. Vet. Diagn. Investig.* **2013**, *25*, 91–98. [CrossRef]
7. Weaver, D.M.; Tyler, J.W.; Marion, R.S.; Wallace, L.M.; Nagy, J.K.; Holle, J.M. Evaluation of Assays for Determination of Passive Transfer Status in Neonatal Llamas and Alpacas. *J. Am. Vet. Med. Assoc.* **2000**, *216*, 559–563. [CrossRef]
8. Carhuapoma Delacruz, V.; Valencia Mamani, N.; Paucar-Chanca, R.; Salas Contreras, W.H.; Morales-Cauti, S. Caracterización de neumonías y bacterias neumónicas en alpacas (*Vicugna pacos*) en comunidades altoandinas de Huancavelica, Perú. *La. Granja* **2022**, *37*, 75–85. [CrossRef]
9. Castilla, D.; Escobar, V.; Ynga, S.; Llanco, L.; Manchego, A.; Lázaro, C.; Navarro, D.; Santos, N.; Rojas, M. Enteric viral infections among domesticated South American camelids: First detection of mammalian orthoreovirus in camelids. *Animals* **2021**, *11*, 1455. [CrossRef] [PubMed]
10. Wellington López, P.; Marycris Chamorro, L.; Antonio, B. Rapid detection of rotavirus and coronavirus in alpaca crias (*Vicugna pacos*) with diarrhea in the Cusco region, Peru. *Rev. Invest. Vet. Perú* **2011**, *22*, 407–411. [CrossRef]
11. Luna, L.; Brandão, P.E.; Maturrano, L.; Rosadio, R.; Silva, F.D.F.; Soares, R.M.; Gregori, F. Betacoronavirus 1 in alpacas (*Vicugna pacos*) in the High Peruvian Andes. *Small Rumin. Res.* **2015**, *133*, 7–9. [CrossRef]
12. Rojas, M.A.; Gonçalves, J.L.S.; Dias, H.G.; Manchego, A.; Santos, N. Identification of two novel rotavirus A genotypes, G35 and P[50], from Peruvian alpaca faeces. *Infect. Genet. Evol.* **2017**, *55*, 71–74. [CrossRef]
13. Cebra, C.K.; Mattson, D.E.; Baker, R.J.; Sonn, R.J.; Dearing, P.L. Potential pathogens in feces from unweaned llamas and alpacas with diarrhea. *J. Am. Vet. Med. Assoc.* **2003**, *223*, 1806–1808. [CrossRef]
14. Cirilo, C.E.; Manchego, S.A.; Rivera, H.H.; Rosadio, A.R. Coexistencia de virus y bacterias en neumonías agudas en alpacas neonatas. *Rev. Invest. Vet. Perú.* **2012**, *23*, 317–335. [CrossRef]
15. Rosadio, R.; Cirilo, E.; Manchego, A.; Rivera, H. Respiratory syncytial and parainfluenza type 3 viruses coexisting with Pasteurella multocida and Mannheimia hemolytica in acute pneumonias of neonatal alpacas. *Small Rumin. Res.* **2011**, *97*, 110–116. [CrossRef]
16. Amer, H.M. Bovine-like coronaviruses in domestic and wild ruminants. *Anim. Health Res. Rev.* **2018**, *19*, 113–124. [CrossRef] [PubMed]

17. Cui, J.; Li, F.; Shi, Z.L. Origin and evolution of pathogenic coronaviruses. *Nat. Rev. Microbiol.* **2019**, *17*, 181–192. [CrossRef] [PubMed]
18. Woo, P.C.; Lau, S.K.; Lam, C.S.; Lau, C.C.; Tsang, A.K.; Lau, J.H.; Bai, R.; Teng, J.L.; Tsang, C.C.; Wang, M.; et al. Discovery of seven novel Mammalian and avian coronaviruses in the genus deltacoronavirus supports bat coronaviruses as the gene source of alphacoronavirus and betacoronavirus and avian coronaviruses as the gene source of gammacoronavirus and deltacoronavirus. *J. Virol.* **2012**, *86*, 3995–4008. [CrossRef]
19. ICTV. International Committee on Taxonomy of Viruses: Coronaviridae. 2023. Available online: https://talk.ictvonline.org/taxonomy/ (accessed on 31 July 2023).
20. Bashor, L.; Gagne, R.B.; Bosco-Lauth, A.M.; Bowen, R.A.; Stenglein, M.; VandeWoude, S. SARS-CoV-2 evolution in animals suggests mechanisms for rapid variant selection. *Proc. Natl. Acad. Sci. USA* **2021**, *118*, e2105253118. [CrossRef]
21. Tao, Y.; Shi, M.; Chommanard, C.; Queen, K.; Zhang, J.; Markotter, W.; Kuzmin, I.V.; Holmes, E.C.; Tong, S. Surveillance of Bat Coronaviruses in Kenya Identifies Relatives of Human Coronaviruses NL63 and 229E and Their Recombination History. *J. Virol.* **2017**, *91*, e01953-16. [CrossRef]
22. Corman, V.M.; Eckerle, I.; Memish, Z.A.; Liljander, A.M.; Dijkman, R.; Jonsdottir, H.; Juma Ngeiywa, K.J.; Kamau, E.; Younan, M.; Al Masri, M.; et al. Link of a ubiquitous human coronavirus to dromedary camels. *Proc. Natl. Acad. Sci. USA* **2016**, *113*, 9864–9869. [CrossRef] [PubMed]
23. Bakkers, M.J.; Lang, Y.; Feitsma, L.J.; Hulswit, R.J.; de Poot, S.A.; van Vliet, A.L.; Margine, I.; de Groot-Mijnes, J.D.; van Kuppeveld, F.J.; Langereis, M.A.; et al. Betacoronavirus Adaptation to Humans Involved Progressive Loss of Hemagglutinin-Esterase Lectin Activity. *Cell. Host Microbe* **2017**, *21*, 356–366. [CrossRef]
24. Vijgen, L.; Keyaerts, E.; Lemey, P.; Maes, P.; Van Reeth, K.; Nauwynck, H.; Pensaert, M.; Van Ranst, M. Evolutionary history of the closely related group 2 coronaviruses: Porcine hemagglutinating encephalomyelitis virus, bovine coronavirus, and human coronavirus OC43. *J. Virol.* **2006**, *80*, 7270–7274. [CrossRef]
25. Falkenberg, S.; Buckley, A.; Laverack, M.; Martins, M.; Palmer, M.V.; Lager, K.; Diel, D.G. Experimental Inoculation of Young Calves with SARS-CoV-2. *Viruses* **2021**, *13*, 441. [CrossRef]
26. Das, T.; Sikdar, S.; Chowdhury, M.H.U.; Nyma, K.J.; Adnan, M. SARS-CoV-2 prevalence in domestic and wildlife animals: A genomic and docking based structural comprehensive review. *Heliyon* **2023**, *9*, e19345. [CrossRef]
27. Hale, V.L.; Dennis, P.M.; McBride, D.S.; Nolting, J.M.; Madden, C.; Huey, D.; Ehrlich, M.; Grieser, J.; Winston, J.; Lombardi, D.; et al. SARS-CoV-2 infection in free-ranging white-tailed deer. *Nature* **2022**, *602*, 481–486. [CrossRef]
28. Fusco, G.; Cardillo, L.; Levante, M.; Brandi, S.; Picazio, G.; Napoletano, M.; Martucciello, A.; Fiorito, F.; De Carlo, E.; de Martinis, C. First serological evidence of SARS-CoV-2 natural infection in small ruminants: Brief report. *Vet. Res. Commun.* **2023**, *47*, 1741–1748. [CrossRef]
29. El Masry, I.; Al Makhladi, S.; Al Abdwany, M.; Al Subhi, A.; Eltahir, H.; Cheng, S.; Peiris, M.; Gardner, E.; Von Dobschuetz, S.; Soumare, B.; et al. Serological evidence of SARS-CoV-2 infection in dromedary camels and domestic bovids in Oman. *Emerg. Microbes Infect.* **2023**, *12*, 2220577. [CrossRef]
30. Ruiz-Arrondo, I.; Portillo, A.; Palomar, A.M.; Santibáñez, S.; Santibáñez, P.; Cervera, C.; Oteo, J.A. Detection of SARS-CoV-2 in pets living with COVID-19 owners diagnosed during the COVID-19 lockdown in Spain: A case of an asymptomatic cat with SARS-CoV-2 in Europe. *Transbound. Emerg. Dis.* **2021**, *68*, 973–976. [CrossRef]
31. Hosie, M.J.; Epifano, I.; Herder, V.; Orton, R.J.; Stevenson, A.; Johnson, N.; MacDonald, E.; Dunbar, D.; McDonald, M.; Howie, F.; et al. Detection of SARS-CoV-2 in respiratory samples from cats in the UK associated with human-to-cat transmission. *Vet. Rec.* **2021**, *188*, e247. [CrossRef]
32. Haake, C.; Cook, S.; Pusterla, N.; Murphy, B. Coronavirus Infections in Companion Animals: Virology, Epidemiology, Clinical and Pathologic Features. *Viruses* **2020**, *12*, 1023. [CrossRef]
33. Vlasova, A.N.; Saif, L.J. Bovine Coronavirus and the Associated Diseases. *Front. Vet. Sci.* **2021**, *8*, 643220. [CrossRef]
34. Savard, C.; Provost, C.; Ariel, O.; Morin, S.; Fredrickson, R.; Gagnon, C.A.; Broes, A.; Wang, L. First report and genomic characterization of a bovine-like coronavirus causing enteric infection in an odd-toed non-ruminant species (Indonesian tapir, *Acrocodia indica*) during an outbreak of winter dysentery in a zoo. *Transbound. Emerg. Dis.* **2022**, *69*, 3056–3065. [CrossRef]
35. Erles, K.; Toomey, C.; Brooks, H.W.; Brownlie, J. Detection of a group 2 coronavirus in dogs with canine infectious respiratory disease. *Virology.* **2003**, *310*, 216–223. [CrossRef]
36. Zhang, X.M.; Herbst, W.; Kousoulas, K.G.; Storz, J. Biological and genetic characterization of a hemagglutinating coronavirus isolated from a diarrhoeic child. *J. Med. Virol.* **1994**, *44*, 152–161. [CrossRef] [PubMed]
37. Rojas, M.; Manchego, A.; Rocha, C.B.; Fornells, L.A.; Silva, R.C.; Mendes, G.S.; Dias, H.G.; Sandoval, N.; Pezo, D.; Santos, N. Outbreak of diarrhea among preweaning alpacas (*Vicugna pacos*) in the Southern Peruvian Highland. *J. Infect. Dev. Ctries.* **2016**, *10*, 269–274. [CrossRef]
38. Rocha, C.B.; Fornells, L.A.A.M.G.; Rojas, M.; Libetal, M.; Manchego, A.; Pezo, D.; Santos, N. Molecular epidemiology of coronavirus in faeces of Brazilian calves and Peruvian camelid herds. *J. Infect. Dev. Ctries.* **2018**, *12*, 37–42. [CrossRef] [PubMed]
39. Jin, L.; Cebra, C.K.; Baker, R.J.; Mattson, D.E.; Cohen, S.A.; Alvarado, D.E.; Rohrmann, G.F. Analysis of the genome sequence of an alpaca coronavirus. *Virology* **2007**, *365*, 198–203. [CrossRef]
40. Crossley, B.M.; Mock, R.E.; Callison, S.A.; Hietala, S.K. Identification and characterization of a novel alpaca respiratory coronavirus most closely related to the human coronavirus 229E. *Viruses* **2012**, *4*, 3689–3700. [CrossRef] [PubMed]

41. Moës, E.; Vijgen, L.; Keyaerts, E.; Zlateva, K.; Li, S.; Maes, P.; Pyrc, K.; Berkhout, B.; van der Hoek, L.; Van Ranst, M. A novel pancoronavirus RT-PCR assay: Frequent detection of human coronavirus NL63 in children hospitalized with respiratory tract infections in Belgium. *BMC Infect. Dis.* **2005**, *5*, 6. [CrossRef]
42. Brandão, P.E. Coronavírus bovino (BCoV): Ocorrência, Diversidade Molecular e Padronização de PCR Para Diagnóstico a Partir de Amostras Fecais de Bezerros com e sem Diarréia Criados em Municípios dos Estados de São Paulo e Minas Gerais, Brasil. Ph.D. Thesis, Universidade de São Paulo, São Paulo, Brazil, 2004.
43. Kumar, S.; Stecher, G.; Li, M.; Knyaz, C.; Tamura, K. MEGA X: Molecular evolutionary genetics analysis across computing platforms. *Mol. Biol. Evol.* **2018**, *35*, 1547–1549. [CrossRef]
44. Lucas, L.J.R.; Morales, C.S.; Barrios, A.M.; Rodríguez, G.J.; Vásquez, C.M.; Lira, M.B.; Torres, L.B.; Casas, A.E.; Espinoza, B.J. Patógenos involucrados en casos fatales de diarrea en crías de alpaca de la sierra central del Perú. *Rev. Invest. Vet. Perú* **2016**, *27*, 159–175. [CrossRef]
45. Martín Espada, M.; Pinto Jiménez, C.; Cid Vázquez, M. Camélidos Sudamericanos: Estado sanitario de sus crías. *Rev. Complut. Cienc. Vet. (RCCV)* **2010**, *4*, 37–50.
46. Islam, A.; Ferdous, J.; Sayeed, M.A.; Islam, S.; Rahman, M.K.; Abedin, J.; Saha, O.; Hassan, M.M.; Shirin, T. Spatial epidemiology and genetic diversity of SARS-CoV-2 and related coronaviruses in domestic and wild animals. *PLoS ONE* **2021**, *16*, e0260635. [CrossRef]
47. Corman, V.M.; Baldwin, H.J.; Tateno, A.F.; Zerbinati, R.M.; Annan, A.; Owusu, M.; Nkrumah, E.E.; Maganga, G.D.; Oppong, S.; Adu-Sarkodie, Y.; et al. Evidence for an ancestral association of human coronavirus 229E with bats. *J. Virol.* **2015**, *89*, 11858–11870. [CrossRef]
48. Anthony, S.J.; Johnson, C.K.; Greig, D.J.; Kramer, S.; Che, X.; Wells, H.; Hicks, A.L.; Joly, D.O.; Wolfe, N.D.; Daszak, P.; et al. Global patterns in coronavirus diversity. *Virus Evol.* **2017**, *3*, vex012. [CrossRef]
49. Bueno, L.M.; Rizotto, L.S.; Viana, A.O.; Silva, L.M.N.; de Moraes, M.V.D.S.; Benassi, J.C.; Scagion, G.P.; Dorlass, E.G.; Lopes, B.L.T.; Cunha, I.N.; et al. High genetic diversity of alphacoronaviruses in bat species (Mammalia: Chiroptera) from the atlantic forest in Brazil. *Transbound. Emerg. Dis.* **2022**, *69*, e2863–e2875. [CrossRef]
50. Chidoti, V.; De Nys, H.; Pinarello, V.; Mashura, G.; Missé, D.; Guerrini, L.; Pfukenyi, D.; Cappelle, J.; Chiweshe, N.; Ayouba, A.; et al. Longitudinal survey of coronavirus circulation and diversity in insectivorous bat colonies in Zimbabwe. *Viruses* **2022**, *14*, 781. [CrossRef]
51. So, R.T.Y.; Chu, D.K.W.; Miguel, E.; Perera, R.A.P.M.; Oladipo, J.O.; Fassi-Fihri, O.; Aylet, G.; Ko, R.L.W.; Zhou, Z.; Cheng, M.-S.; et al. camel coronavirus HKU23 in African camels revealed multiple recombination events among closely related Betacoronaviruses of the subgenus Embecovirus. *J. Virol.* **2019**, *93*, e01236-19. [CrossRef]
52. Su, S.; Wong, G.; Shi, W.; Liu, J.; Lai, A.C.K.; Zhou, J.; Liu, W.; Bi, Y.; Gao, G.F. Epidemiology, genetic recombination, and pathogenesis of coronaviruses. *Trends Microbiol.* **2016**, *24*, 490–502. [CrossRef]
53. Quintana, H.; Pacheco, V. Identificación y Distribución de los Murciélagos Vampiros del Perú; Lima. *Rev. Perú. Med. Exp. Salud Publica* **2007**, *24*, 81–88. Available online: http://www.scielo.org.pe/scielo.php?script=sci_arttext&pid=S1726-46342007000100011 (accessed on 1 August 2023).
54. Bergner, L.M.; Orton, R.J.; da Silva Filipe, A.; Shaw, A.E.; Becker, D.J.; Tello, C.; Biek, R.; Streicker, D.G. Using noninvasive metagenomics to characterize viral communities from wildlife. *Mol. Ecol. Resour.* **2019**, *19*, 128–143. [CrossRef]
55. Simas, P.V.M.; Barnabé, A.C.S.; Durães-Carvalho, R.; Neto, D.F.L.; Caserta, L.C.; Artacho, L.; Jacomassa, F.A.F.; Martini, M.C.; Bianchi dos Santos, M.M.A.; Felippe, P.A.N.; et al. Bat coronavirus in Brazil related to appalachian ridge and porcine epidemic diarrhea viruses. *Emerg. Infect. Dis.* **2015**, *21*, 729–731. [CrossRef]
56. Ahmad, T.; Khan, M.; Haroon; Musa, T.H.; Nasir, S.; Hui, J.; Bonilla-Aldana, D.K.; Rodriguez-Morales, A.J. COVID-19: Zoonotic Aspects. *Travel Med. Infect. Dis.* **2020**, *36*, 101607. [CrossRef]

MDPI AG
Grosspeteranlage 5
4052 Basel
Switzerland
Tel.: +41 61 683 77 34

Animals Editorial Office
E-mail: animals@mdpi.com
www.mdpi.com/journal/animals

www.ingramcontent.com/pod-product-compliance
Lightning Source LLC
Chambersburg PA
CBHW062039220926
51850CB00010B/2608